Tanmay Teaches Go

Tanmay Teaches Go

The Ideal Language for Backend Developers

Tanmay Bakshi
Baheer Kamal

New York   Chicago   San Francisco
Athens   London   Madrid
Mexico City   Milan   New Delhi
Singapore   Sydney   Toronto

Library of Congress Control Number: 2021935615

McGraw Hill books are available at special quantity discounts to use as premiums and sales promotions or for use in corporate training programs. To contact a representative, please visit the Contact Us page at www.mhprofessional.com.

Tanmay Teaches Go:
The Ideal Language for Backend Developers

1 2 3 4 5 6 7 8 9   LCR   26 25 24 23 22 21

ISBN   978-1-264-25814-7
MHID   1-264-25814-3

This book is printed on acid-free paper.

Sponsoring Editor Lara Zoble	**Copy Editor** Michael McGee
Editing Supervisor Stephen M. Smith	**Proofreader** Claire Splan
Production Supervisor Lynn M. Messina	**Indexer** Claire Splan
Acquisitions Coordinator Elizabeth M. Houde	**Art Director, Cover** Jeff Weeks
Project Manager Patricia Wallenburg, TypeWriting	**Composition** TypeWriting

About the Authors

Best known as the youngest artificial intelligence (AI) expert, 17-year-old **Tanmay Bakshi** is the co-author of the book that you're reading now under the *Tanmay Teaches* series. He started playing and working with tech at the age of 5 and today is the bestselling author of his first book *Hello Swift!* and of *Tanmay Teaches Julia for Beginners,* and co-author of *Cognitive Computing with IBM Watson.*

Tanmay is a TED and keynote speaker, Top 25 Tech Influencer on LinkedIn according to Onalytica, Google Developer Expert for Machine Learning, and IBM Developer Advocate, and the host of the weekly live show called Tech Life Skills with Tanmay on his YouTube channel, Tanmay Teaches, with over 300,000 subscribers and 4.5 million views.

He's had the privilege of keynoting on the world's largest stages for major international organizations, such as the United Nations, Apple, Google, Microsoft, IBM, Walmart, SAP, and more. He's also taught at and hosted workshops for major schools and universities across the globe.

Tanmay is a media personality and his work has been featured in *Forbes, The Wall Street Journal, The New York Times,* CNBC, and *Bloomberg Businessweek,* just to name a few. He builds technology that augments human lives, and aspires to empower at least 100,000 people with the technology they need to solve the problems they see in the world. Therefore, he's had the honor of being the recipient of the Life Mentor Award from the Creative Foundation, Knowledge

Ambassador Award from the Sheikh Mohammed Bin Rashid Al Maktoum Knowledge Foundation, IBM Champion 2016–2019, *Toronto Star* Newsmaker of the Year 2013, and Twilio Doer Award.

Baheer Kamal is an Advisory Software Engineer at IBM. By the age of 13, he taught himself Java, including advanced functionality like parallelism. Since then, he's been coding in languages such as Go, Bash, Python, and more. At IBM, he's responsible for developing and maintaining large codebases in these languages.

Contents

Preface ix

Acknowledgments xi

1 Introduction 1

Why Go? 2

    What Are Go's Design Goals? 2

    Where Can I Use Go? 3

    The Go Compiler and Runtime 4

    Concurrency 6

2 Quickstart 9

Installing Go 10

    Using a System Package Manager 10

    Manually Installing a Precompiled Binary 10

    Building Go from Source 11

Basic Go Concepts 12

    Conventional Project Structure 12

    A Go Primer—"Hello, World!" 14

    Variables 16

    if Statements and switch Statements 23

    Loops 26

    Functions 31

Structures 42
Interfaces 47
Errors 51

3   **Go Modules** **59**
Using Built-in Packages 60
Using Third-Party Packages 71
Am I Prime? 71
Building Your Own Packages 74
Am I Prime? Part 2 74

4   **Using Built-in Packages** **81**
Common Data Structures and Algorithms 82
Dijkstra's Pathfinding 82
Conway's Game of Life 99
Proof of Work 112

5   **Concurrency** **123**
Concurrency, Threads, and Parallelism 124
Goroutines 126
Channels 129
select Statements 139
Proof of Work: Part 2! 144

6   **Interoperability** **155**
Why Is Interoperability Important? 156
Interoperating with C Code 159
Interoperating with Swift 169

Index.. 183

Preface

*T*anmay Teaches Go is intended to make the Go programming language easily accessible and approachable for every developer. Go, as a language, is built with modern backend development in mind. As Moore's Law comes to an end, it's no longer enough to expect better and faster computers. Therefore, we must adapt to the new normal: large-scale, distributed systems, running highly optimized software stacks.

These distributed systems, with their fundamentally different hardware ideology, have also forced the software world to innovate and change as well. Modern software systems are moving away from large, monolithic designs. Rather, they're powered by lots of smaller *microservices*, enabling greater reuse of code across projects, and a delegation of responsibilities—all of which results in cleaner code with smaller surface area for bugs.

As the software world has had to evolve, the demand for the right infrastructure for this new paradigm has grown, resulting in brand new *programming languages*, developed from the ground up, to meet that demand. Go is one of these languages, having specific traits suitable for microservices-based software development, such as fast compile times, small binary sizes, and portability. There are also special features that no other language can really match, such as Goroutines and channels, for powerful, easy-to-use concurrency.

Go is the engine behind many of the infrastructural services in backend development, such as Kubernetes, Docker, etcd, CockroachDB, and more. By using the same language that these popular tools are already built in, your code is closer to, and can take better advantage of, the large ecosystem of open source code meant for enterprise-level backend development.

Through this book, we hope to turn you into *Go developers*, so you have the capability to leverage Go as a tool in your problem-solving toolbox. The best part is that Go isn't just any other tool. It's a tool that can be applied to solve *enterprise-level* problems; therefore, it's a prime candidate to be your first choice to write code in.

Our goal is not only to teach you the Go language, but take an approach that, at every step, contrasts and compares Go to other popular programming languages used for similar tasks, such as Swift, C, Python, and Java. We even go down to the operating system and CPU level to show you exactly how these languages interact with your computer's hardware and software. The purpose for this is twofold:

- We want to make it easy for developers coming from other languages to transition to the world of Go.
- We want you to derive insight from the decisions made by other programming languages and the low-level architecture of our computers. This helps you better understand which tasks Go is and is not well-suited for, as well as how it works internally and why it works that way.

Tanmay Bakshi
Baheer Kamal

Acknowledgments

I'd like to say a huge thank you to my family and my mentors for their unconditional support, which keeps me motivated. Special thanks to the wonderful team at McGraw Hill for supporting me as I wrote the second of many books to come in the *Tanmay Teaches* series. I'd especially like to thank Lara Zoble at McGraw Hill for providing me with the support, advice, and feedback I needed to make this a success. She is the brain behind the creation of the *Tanmay Teaches* series.

The idea of authoring a book on Go originated when my co-author, Baheer, and I were writing code for a project. We both had the same objective: to sift through and massage large binary files. I implemented in Swift, and he implemented in Go. On the very first run, Go blew Swift out of the water in both compile-time and run-time benchmarks! Not only that, the code was more intuitive. This inspired me to write a book on the Go language, its strengths, why it works, and where to apply it. My special thanks to Baheer for being with me on this journey!

Tanmay Bakshi

Tanmay Teaches Go

CHAPTER 1

Introduction

Welcome to *Tanmay Teaches Go*! In this book, the second in the Tanmay Teaches series, you will learn about the world of Go programming. Go is a programming language designed with safety, performance, flexibility, and *concurrency* in mind. Its design goals fundamentally align with the needs of backend development, where Go has found the most value.

Once you've finished this chapter, you'll be able to answer the following questions:

- What are Go's design goals that make the Go language special, and what are some of its drawbacks?
- What platforms does Go support, and where can it be used?
- Why is Go's compiler so special?
- What are Go's memory management and its Garbage Collector?
- How does concurrency make the Go runtime so unique?

Why Go?

Go is a language built with a very specific purpose in mind: to ease the entire workflow of a backend developer—all the way from speed of execution, speed of compilation, and dependency management, to concurrency and optimization. To achieve this vision, the Go team has built the language spec with a few key design goals in mind.

What Are Go's Design Goals?

Go was developed by Google back in 2009, and the first stable version was released to the public in 2011. The language was designed by Robert Griesemer, Rob Pike, and Ken Thompson. Go is known for its mascot: the Gopher, as shown in Figure 1.1!

FIGURE 1.1 "Gopher": The Go mascot

Go emerged as a solution to a number of problems that Google was facing. On the compilation front, for example, the majority of compilers were too slow to compile Google's large codebases in a reasonable amount of time.

On the implementation front, languages such as C++ proved to be too low level and verbose, and languages such as Java proved to be too high level and heavyweight (with the baggage of the JVM and the latency introduced by the Garbage Collector).

Google decided the best way to solve these problems was to, well, develop their own language!

Google's design goals for Go were as follows:

- Create a compiler that works on large codebases as fast as possible
- Create a compiler that generates fast code using a good set of optimizations

Go delivers on exactly those goals. The compiler has proven to be very fast. To enforce this kind of speed, the compiler inherits a few key traits:

- It's incredibly strict with your code; unused variables and imports, for example, are *errors*, not *warnings*.
- The compiler architecture is *simple*. The majority optimizations, for example, take place on the compiled assembly code, not on an internal intermediate representation of the Go code (as in Swift, C, C++ compiling through LLVM).
- Assembly is generated directly in binary form, instead of being generated as text and compiled to binary.

Not only is the compiler itself fast, it's also able to generate fast-running code because of the way the language is structured. By keeping language features and syntactic sugar slim, the optimizer within the compiler has less work to do analyzing complex control flow and logic. This means that Go code can be quite terse at times, but the result is fast compiled code that you get quickly.

At the same time, due to the strictness of the compiler, you also inherit safety. In more complex languages that are known for their best in class safety features, such as Swift, implementing that safety requires quite a bit of extra overhead in the compiler and some overhead in the runtime. However, because Go syntax is very minimal, safety is simple to implement in both compile-time and runtime.

Where Can I Use Go?

Go is also *highly compatible* across all sorts of platforms! For example, it supports:

- Linux
- BSDs (Berkeley Software Distributions)
  - DragonFly BSD/FreeBSD/OpenBSD/NetBSD, etc.
  - Darwin
    - macOS/iOS/watchOS/tvOS, etc.
- Plan 9
- IBM i
- Solaris
- Windows

As shown in the preceding list, Go even supports Plan 9 and IBM i! What more could you ask for?

Not only does Go support a variety of operating systems, it also supports a range of CPU architectures, such as:

- **x86/x86_64**—The most mainstream CPU architecture today.
- **ARM/ARM64**—An architecture built for mobile systems, which will be mainstream in high-performance computing soon, thanks to Apple Silicon.
- **MIPS**—A very simple ISA that's uncommon in the real world.
- **IBM PowerPC**—An enterprise-grade RISC CPU architecture by IBM that values high-speed I/O with accelerators, performance, and massive parallelism.
- **IBM Z**—A family of enterprise-grade CISC CPUs (in *mainframes*!) by IBM that values extreme redundancy, reliability, and performance.
- **RISC-V**—An incredibly exotic, nearly brand new CPU architecture for which there's *only one chip* available in the market today, by SiFive.

If Go even supports RISC-V and IBM Z, I'm sure it'll fit your use case.

The Go Compiler and Runtime

A common criticism of Go is precisely one of the things it was built to have—a very simplistic syntax. (This syntax style is also, at times, inconsistent, but this is a topic for another time.) In fact, there's a famous joke that the Go language designers ignored all programming language technologies invented after 1980! Despite this less modern way of representing code, however, Go powers some of the most modern software, such as Docker, Kubernetes, Terraform, CockroachDB, etcd, and more.

One key reason Go is so pervasive is its performance. We've already talked about how the Go compiler is smart, modern, and fast, but the way the language is designed enables programmers themselves to write fast code, too. For example, you can get down to the level of C, dealing with pointers and manually managing memory, or you can abstract all of that away and let Go deal with memory for you.

Memory Management and the Special Garbage Collector

Specifically, Golang uses Garbage Collection (GC) for memory management. There are four mainstream ways to do memory management in programming languages:

- **Don't!**—This is the approach taken by C, in which you do *everything* manually.
- **Garbage Collection (GC)**—This is the most popular approach, by far, but unfortunately the one that usually compromises the most when it comes to program performance (of course, there are exceptions to every rule, and this is just the average case). For example, Java and Kotlin (via the JVM), Python, C#, Go, JavaScript, Lisp, Ruby, and Julia all use Garbage Collection. The way this works is that, every so often, the "Garbage Collector" kicks in and looks for memory that no longer has any references to it. If there are no references, that means the program is no longer using it, and it can be deleted. The issue with this method, or at least most of its implementations, is that it introduces short "halts" into the program when the app has to stop and collect garbage. It's also not deterministic, so you can't be absolutely certain of—at any point in your application—how much memory you really have, without explicitly checking. You can't even predict when your app will next come to a halt due to Garbage Collection.
- **Automatic Reference Counting (ARC)**—This method is usually the best trade-off between performance, smoothness (no jittering), and ease of implementation. With this memory management technique, every time you create a new object, you also create a "reference counter" alongside it. Every time you pass a reference to the object somewhere, the counter increments. Every time a reference goes out of scope, the counter decrements. Every time the counter is decremented, a check is run to see if the counter is at 0. If it is, the memory is freed, because the last reference has gone out of scope. Languages such as Swift and Objective-C use this method. However, you sometimes need to be careful about *reference cycles*, which can lead to memory never getting freed, although this is easy to find with some simple instrumentation tooling and is easy to fix with *weak* and *unowning* variables.
- **Compile-Time Ownership Disambiguation**—As the name suggests, this method is the one that really values performance and safety, while also being slightly more convenient than just leaving memory manage-

ment as something that needs to be done entirely manually. Languages such as Rust use this method. This technique makes it so the compiler can determine, as it compiles the code, when memory will need to be allocated or freed. This can be a bit complex to code for because there are more restrictions the compiler must place to disambiguate the ownership of memory. However, it is worth it for system programming languages such as Rust.

Online, you'll find many varying opinions about which memory management technique is best, but it really boils down to the implementation. For example, some corner cases, such as Intel's open-source user-space network driver, are architected in a way to be optimized for a specific kind of memory management; this makes languages such as Swift that use ARC very slow. In the common case, it's GC that's usually the slowest, most jittery, or most inconvenient. However, that's just the common case—some implementations can be exceptional, such as the GC in Go!

Go's Garbage Collector is so efficient, in fact, that there are usually no noticeable jitters in between program execution, and it can even match the performance of Swift's ARC.

Concurrency

We've really only just gotten started talking about the advantages of the Go language, but there is one more flagship feature, which may just be the tipping point in convincing you to use Go: *easy, powerful concurrency.*

Moore's law is slowly but surely coming to an end. That's sad, but we have to do something about it. So far, our best solution is multiprocessing, which is when a single chip can execute multiple sets of instructions at the same time, in parallel. However, driving so many "cores" can be difficult, because you need to make sure you're saturating them with work, and they're not just sitting around waiting for system calls, memory accesses, etc.

To make multiprocessing easier, operating systems have a concept of "threads." On most platforms that are POSIX-compliant, threads are part of a "process" and have their own stack pointers. On other systems, such as Windows, threads are their own processes. You can have hundreds of threads running through your kernel, but the problem is that these are relatively heavyweight. For example, on a 22-core system with 44 threads, you can scale to "thousands" of OS threads on Linux. However, when you get to a larger scale, the pure cost of the "Linux CFS (Completely Fair Scheduler)" simply context switching between the threads becomes too much to handle.

In order to fix this problem, Go introduces the concept of Goroutines. A Goroutine is a super lightweight "thread-like" execution environment, which is handled by the Go runtime rather than the OS. Goroutines map to OS threads, which then further map to CPU threads, enabling millions of Goroutines to run without sacrificing much performance.

Now, I know what you're thinking: "These must be a pain to use!" The truth is quite the opposite, actually. It's easier to implement Goroutines in Go than it is to use Jobs on Grand Central Dispatch in Swift! It's quite literally just a single token of syntax—"go"—that tells Go to execute a function in a separate Goroutine.

And, to make communication among Goroutines high performance, Go features "channels." Channels make it super-easy and fast for Goroutines to communicate with one another.

These two features are highly sought after in other languages, such as Swift and C. There are actually quite a few open-source implementations of Goroutines and Channels in other languages, but none can really match the original!

The fact that Go is the backend for popular services such as Docker and Kubernetes, and that Go can scale with Goroutines, plus the fact that it supports so many architectures and is so lightweight, makes it the perfect language in which to architect the microservices that power large, complex services.

Finally, there are a couple of more plus points to Go that aren't *unique* per se, but they make coding in Go more convenient than in other languages:

- Not only does Go support tons of architectures and operating systems, you can also cross-compile for any architecture, from any architecture, by just passing a couple of extra compile flags to Go! No extra libraries need to be installed. Try doing that with Clang! On Ubuntu, you'd be stuck in Aptitude dependency hell for at least tens of minutes, if not hours.
- Go features an experimental LLVM backend, enabling you to compile Go code to LLVM in case your use case can benefit from the optimizations or flexibility of the modular LLVM compiler.

The preceding text has been a sneak peek into the world of Go and what makes it so special, but we've only just gotten started. Next, in Chapter 2, let's install Go, take a look at some basic Go constructs, get you calibrated to Go standards and syntax, and dive into building some apps!

Exercises

1. Who created Go and for what purposes?
2. In your view, what are the three main factors that would make Go your programming language of choice?
3. What are the four most common ways that programming languages allocate and reclaim memory, and what are their pros and cons?
4. What are some of the drawbacks of Go?
5. Why is the Go runtime so unique?

CHAPTER 2

Quickstart

L et's get started! But before we get to the fun part, which is writing code, it's important to make sure we have an environment in which we can write Go code, compile it, and run it as well.

Once we have our environment set up, we'll take a look at some of the basic building blocks of the Go language, such as the different kinds of control flow, and how they work together.

Once you've finished this chapter, you'll be able to answer the following questions:

- How do you install Go on a Windows, Linux, or macOS machine?
- How can you find out how to install Go from source?
- How does basic control flow work in Go?

Installing Go

As with most languages, installing Go on your system is as simple as running a few system commands. We will explore three ways of installing Go:

- Using a system package manager
- Manually installing a prebuilt binary
- Building Go from source

Using a System Package Manager

The easiest way to install Go, much like any other program, is to use the package manager built into your system. Unfortunately, Windows doesn't come with a full-featured built-in package manager, which is one of the reasons it may not be an optimal system for backend developers. On Linux and macOS, however, you can use the package managers and commands explained in Table 2.1.

TABLE 2.1   Commands to Install Go Using Built-in Package Managers

OS/PACKAGE MANAGER	INSTALLATION COMMANDS
Ubuntu/Aptitude	`sudo snap install golang --classic`
macOS/Homebrew	`brew install go`

Unfortunately, the package manager distributions are only *community supported*, and not officially supported by the Go team. While this method is more convenient, for real production workloads, which may require a specific version of Go, for example, we recommend downloading a precompiled binary.

Manually Installing a Precompiled Binary

The exact process of installing a precompiled binary depends on the platform you're running on. We'll take a look at Ubuntu and macOS as examples, but if you want to see instructions specific to your platform, please visit the Go website at https://golang.org/doc/install.

Installing a precompiled binary boils down to a few key steps:

1. Downloading the binary
2. Extracting the binary
3. Updating your PATH so your shell can find the binary

You can download Go and its accompanying tools from https://golang.org/dl. For Linux, after downloading, you'll have a ".tar.gz" file. One of the ways you can extract it is from the command line:

```
tar -xvf go1.14.6-linux-amd64.tar.gz
```

NOTE: The version "1.14.6" that we've hardcoded into the preceding command will need to be replaced based on the version you downloaded or the name of your archive.

This will give you a "go" folder, under which is a "bin" folder, which contains the "go" binary—the one tool you need for everything Go related! However, manually invoking that Go binary with an absolute path every time is very inconvenient. Therefore, when you extract, use this command instead:

```
tar -C /usr/local -xvf go1.14.6-linux-amd64.tar.gz
```

This will deflate and move the "go" folder in your "/usr/local" directory. Next, add the following line to your "$HOME/.profile" file:

```
export PATH=$PATH:/usr/local/go/bin
```

From here on out, you will be able to access the "go" binary just by typing, well, "go", in your terminal!

Building Go from Source

Building the Go compiler and tools from source is a lot more intense. One reason for this is because Go is written in Go itself, which introduces a sort of chicken-and-the-egg problem. Thus, just as with a C compiler, you have to "bootstrap" a compilation of the Go compiler using an older Go compiler.

NOTE: To give you a sense of the performance of Go: The Go compiler is known to be one of the fastest code compilers in the world. What language is it written in? Go! It just goes to show that the Go language developers practice what they preach.

Technically, there is another way to compile Go without using an older version of the official Go compiler: Use a different compiler that implements the same Go language specification. When we refer to "Go", we usually mean "gc", which stands for "Go compiler," but it really means the "Go language spec." For example, before the first version of "gc" was written, you'd have to use "gccgo", which is a GCC (GNU C Compiler) frontend for the Go language. This frontend is still maintained (albeit a few versions older than gc), and can be used to compile gc itself.

To this day, gc has more of a focus on compiling code *fast*, and while it does generally output very high-quality machine code, there are some cases when tight loops or performance-sensitive code may need to be compiled with a compiler that takes its time and optimizes better. For example, gc doesn't support tail call optimization, making some recursive algorithms in Go run slower. You can use gccgo for these kinds of codebases.

To learn how to compile Go from source, visit their GitHub Wiki page at https://github.com/golang/go/wiki/InstallFromSource.

Basic Go Concepts

Now it's time to dive into actually using some of the core concepts you'll use every day as a Go developer. These core concepts include basic control flow, project structure, syntax for more advanced features such as functions and structures, and also error handling. Let's get started!

Conventional Project Structure

Unlike some other languages, Go doesn't have a very strict way of laying out your project. However, there are standards that the majority of projects conform to that make it easier for developers to understand the way code bases are structured and what the code does.

You can find out more about this official standard at the following GitHub repository at https://github.com/golang-standards/project-layout.

Before we can talk about project layout, there is an important piece of terminology you need to understand: *packages*. A Go package is a single source code file. The standard for these packages is that the package name for a file is the same as the filename itself, without the file extension.

For example, for source file "main.go", the package name is "main", which is declared in the file's first line as the following valid Go syntax:

```
package main
```

Returning to directory structure . . . In brief, Table 2.2 represents the directory structure you need to know for now. In the next few chapters, as we cover more functionality, we'll talk about more directories you can use in your projects.

TABLE 2.2   The Purpose of the Files and Folders in the Default Go Project Structure

PATH WITH RESPECT TO PROJECT ROOT	PURPOSE
/vendor	This directory stores the dependencies of your project. It can be created automatically by running "go mod vendor". Certain IDEs (Integrated Development Environments), such as IntelliJ, can integrate with this directory to provide dependency management capabilities.
/internal	This directory is for the code that you *don't want people to use*. For example, if you're working on a library, and you want a certain set of APIs (application programming interfaces) to remain private (such that the users of the library can't call those APIs), the standard is to place that code in this directory. You can have multiple "internal" directories in your program, within different modules.
go.mod	This is not a directory; it is a file at the top level of your project folder. This file, known as the "Go module file," stores information that remains the same across the "packages" in your code. For example, the name of the module that contains the packages you've created, the version of Go used to compile the packages, and the dependencies to grab for the code along with their versions.
go.sum	This is, once again, a file. This one isn't as important for you to understand at this stage. It contains the cryptographic hashes of the content of specific module versions. This can help Go detect when a certain dependency is corrupted or has any unexpected content.

Now that you know a bit more about how Go code is structured, let's start learning some syntax!

The basic building block of the syntax of any programming language is the "tokens" you use to build expressions. Some tokens are reserved as "keywords" in this language, meaning you cannot use them as the name of a variable or function, for example. That's because these tokens are integral to the way the Go compiler actually parses your code to know which parts mean what. Table 2.3 enlists all of those keywords.

TABLE 2.3 List of Keywords Reserved in Go

break	struct	range
default	chan	type
func	else	continue
interface	goto	for
select	package	import
case	switch	return
defer	const	var
go	fallthrough	
map	if	

Now, let's start experimenting with some Go syntax!

A Go Primer—"Hello, World!"

To start, let's take a look at a simple program that gives you a good feel for Go. So, as is programming tradition, let's implement "Hello, World!"

Begin by creating a new file called "main.go" in a new directory called "helloworld". Your directory structure should look something like this:

```
/helloworld
    main.go
```

Within the main.go file, enter the following code.

CODE LISTING 2.1 The traditional "Hello, World!" in Go

```
package main

import (
    "fmt"
)

func main() {
    fmt.Println("Hello, World!")
}
```

That's all there is to it! If you navigate to your terminal, go into your helloworld directory and run:

```
go build
```

You should now have an executable called "helloworld", in the same directory. In other words, you've just created and compiled your very first Go app, which you can now run! On a UNIX-like system, you can invoke (or "run") it like so:

```
./helloworld
```

You should then see the following output:

```
Hello, World!
```

The program does work—but how? Let's go back and analyze the source code part by part, as you can see in Table 2.4.

TABLE 2.4 Functionality of Each Section of the "Hello, World!" Code

SOURCE SNIPPET	PURPOSE
`package main`	As mentioned before, Go code is organized into "packages"—each file representing a single package. This line tells Go that the "main.go" file contains a package called "main".

`import (`
`    "fmt"`
`)`
	Go's built-in function for printing to `stdout` is very limited and not used all that often. So, instead, we import a module with a more powerful printing function. This module is built in, and it's called "fmt", which stands for "Format." This code will help us import "fmt". Later, we'll see the syntax for importing multiple modules.

`func main() {`
`    fmt.Println("Hello, World!")`
`}`
	This is the "main" function, which acts as the entry point of the application—a classic choice! Within it, we only execute a single expression. We tell Go: "from within the 'fmt' module, find the 'Println' function, then call it, and pass it the following string literal as the only argument: 'Hello, World!'"

NOTE: One difference you may have noticed about Go syntax is that the first letter of the `Println` function is, well, a capital letter. More on that a bit later. There is quite an intriguing reason for this.

Variables

The Go language is both strongly and statically typed. That means two things:

1. Variables are not automatically converted from one type to another by the compiler in a "hidden" fashion. Even casts such as 32- to 64-bit integers must be completed manually.
2. The compiler *can* try to figure out the type of variables at compile time, instead of having you specify the type yourself. However, Go doesn't do "Type Inference" in as advanced a manner as Swift does—again, that's because the compiler wants to be *as fast* as possible. We'll explore some of these limitations as we progress.

The standard in Go is to allow the compiler to figure out the type and to avoid manual annotation, unless it's necessary, such as to tell the compiler not to use a type that translates directly from a certain literal value (e.g., "for this number, please don't use the default 64-bit integer, use an 8-bit integer instead"). You will see multiple examples of this manual annotation as we continue.

Let's start exploring the world of variables by looking at global variables. A global variable can be one of two types: static or constant. A static global variable can be declared like so:

```
var age = 16
```

That expression does two things, not just one. It first declares a new global static variable called "age", and then initializes that variable with the value 16. In this case, the initialization and declaration are able to work together to mark this variable as an "int" type. However, if you are unable to initialize the value, and just want to "declare" that a global static variable exists, you must explicitly code the type of the variable yourself, like so:

```
var age int
```

If you'd like, even though it's not considered good practice, you can combine the approaches and explicitly declare the type of the variable and provide a value with which to initialize it:

```
var age int = 16
```

Because it's a global *static* variable, it's allowed to change. So, if you have a program such as the following:

CODE LISTING 2.2   Getting and setting a global static variable

```go
package main

import (
    "fmt"
)

var age = 16

func main() {
    fmt.Println(age)
    age = 24
    fmt.Println(age)
}
```

You should see the following output:

```
16
24
```

Now, as a heads up, in a lot of Go programs, you'll see a lot of variable, function, and structure names that start with an uppercase letter (and follow the UpperCamelCase standard), instead of the more traditional lowercase one. For example, `var Age int` instead of `var age`. Why is this?

It's because the capitalization of the first letter is how *Go enforces access control*. In other languages, such as Swift and Java, you must annotate your symbols (for example, function and variable names) with "public," "private," and other access control keywords to get the compiler to enforce where, for example, certain variables and functions can and cannot be reached. In Go, however, there are no annotations for access control. The capitalization of the first letter itself determines if your variable is `public` or `private`. If it's capitalized, it's `public`; if it's lowercase, it's `private`!

NOTE: This decision is definitely a controversial one, because deciding whether or not you want your code to be accessible publicly is now tied to the stylistic choice of your code.

Many languages like to enforce access control in their own ways, so Go isn't incredibly different in that regard. Python, for example, encourages duck typing, which means it doesn't enforce those rules *at all!*

One more thing to know about syntax related to variables: If you'd like to initialize multiple variables at once, you don't need to repeat the "var" keyword multiple times. You can simply write:

```
var (
    name1 = "Tanmay Bakshi"
    name2 = "Baheer Kamal"
)
```

This will declare and initialize those two strings as global variables. You can even mix and match types like so:

```
var (
    name1 = "Tanmay Bakshi"
    age = 16
)
```

Notice the lack of commas at the end of each line—the value assignment itself acts as the delimiter between variables.

Now that you know how to work with `global static` variables, let's take a look at global constants. Constants, as the name implies, strictly *cannot be modified*. In languages such as Go and Swift, where safety is valued immensely, you can't even do tricks such as "get a pointer to the constant and modify it that way," because the compiler won't let you. This makes it so the compiler can be absolutely certain when running optimizations, such as constant folding, which precomputes some calculations to avoid executing them at runtime.

A global constant is created by replacing the word "`var`" in a variable declaration with "`const`". It can be used like so:

```
const name1 = "Tanmay Bakshi"
const name2 string = "Baheer Kamal"
```

Both of these are valid, but the first one relies on type inference, whereas the second relies on a manually annotated type, which in this case is `string`, for the Go compiler. Keep in mind that while a variable can be declared but may not be initialized, a global constant cannot *just* be declared, because it cannot be changed in the future.

For some types, such as maps, the Go compiler cannot ensure that you won't use some trickery in your code to mutate the internals of the constants. Therefore, you're not allowed to store those types inside of constants; you must use a global static variable instead.

All right, now you know about global variables. So, let's talk about local variables! Creating local variables is done quite differently in Go, compared to other languages, using the infamous `:=` operator. Let's take a look at an example.

If you want to create a local variable inside your main function and print it out, you could do it like the following:

CODE LISTING 2.3   Local variables within functions

```go
package main

import (
    "fmt"
)

func main() {
    name := "Anna"
    fmt.Println(name)
}
```

We know the variable instantiation looks weird when you compare it to other languages, but the first line of code in the main function told Go "create a variable called `name` and store the string literal 'Anna' within it." What *hasn't* changed when you compare this to another language such as C or Swift is scope. A variable is deleted after the block of code it's in exits.

"Deleted" is a bit of a strange concept, so let's dive a bit deeper into that. If you have a float such as the following:

```go
price := 69.99
```

The variable "`price`" takes 4 bytes of memory, and it directly contains those 4 bytes, as in, a CPU register could hold those 4 bytes directly. So, when the block of code it's in exits, in this case the main function, it's deleted, and it no longer exists. However, if the variable is, in reality, a reference to another variable or structure, then the reference is deleted, which will, sometime in the future, trigger the Garbage Collector to remove whatever it was referring to *if and only if* there remain no other references to it as well.

If you'd like to declare a variable locally but not initialize it, you can use the same global static variable syntax:

```
func main() {
    var name string
    name = "Jackson"
    fmt.Println(name)
}
```

You can even create local constants:

```
func main() {
    const pi float32 = 3.14159
    fmt.Println(pi)
}
```

We've already covered some basic types of variables, but there are some more you should keep in mind, which are listed in Table 2.5.

TABLE 2.5   What the Common Native Types in Go Store

GO TYPE	WHAT IT STORES
string	Textual information
int8/int16/int32/int64	n-bit signed integer
uint8/uint16/uint32/uint64	n-bit unsigned integer
uint/int	32 bit (un)signed integer on 32-bit platforms 64-bit (un)signed integer on 64-bit platforms
float32/float64	n-bit floating point

Of course, variables aren't limited to being scalar, you can create arrays as well. Creating an array is quite simple in Go. In terms of a type annotation, it's just appending the type you want in the array to the [ ] token. For example, for an array of strings, you'd use the [ ]string type annotation.

An array literal's elements are wrapped in curly braces instead of square brackets, and the array's type annotation must always precede the values within the literal. For example, if you want to create a local array of strings, you could use the following:

```
names := []string{"Tanmay Bakshi", "Baheer Kamal"}
```

Or, if you want to manually annotate the type, you could use this:

```
var names []string
names = []string{"Tanmay Bakshi", "Baheer Kamal"}
```

In order to append a new element to an array, use the "append" function:

```
names = append(names, "Kathy")
```

The append function allocates memory for a new element at the end of the array, and then sets the value of that new space in memory to the second argument we passed to the function. It then returns a NEW array, which we place back into the original variable.

An array in Go is stored as a contiguous block of memory, meaning that, from a pointer to the first element, you can reach the rest of the elements in subsequent memory addresses.

Internally, when you append to an array, Go's runtime may need to run an operation known as a "reallocation." Most operating systems try to extend the buffer of memory that you already have, but when that's not possible because the succeeding memory is being used by something else, then, to accommodate the bigger array, it allocates a brand new buffer, copies the previous elements over, and then frees the previous buffer.

As you can tell, that whole mechanism can be quite slow. In order to make some array operations more efficient when we already know *how many* elements we'll need, but not what those elements are, you can use the "make" function. With the "make" function, you can allocate a certain number of elements in the array for you to use without having to run the reallocation. Here's an example of that:

```
names := make([]string, 3)
names[0] = "Tanmay Bakshi"
names[1] = "Baheer Kamal"
names[2] = "Kathy"
```

One thing to keep in mind is that this does not only reserve capacity in the array, it also reports that the length of the array *is* the number of elements you passed into "make". So, if you make an array with three elements, and don't set their values, the memory within them is just garbage memory from whatever was last stored there.

In some other languages, such as Swift, when you reserve capacity, it still reports that the array is empty, and when you append, it simply doesn't need to do any reallocations until you reach that reserved capacity count. To achieve the same outcome in Go, of an empty array with a backing memory buffer with a certain capacity, you can pass the make function a zero and then pass it the capacity you'd like as a third argument. For example:

```go
func main() {
    names := make([]string, 0, 2)
    fmt.Println(len(names))
    names = append(names, "Tanmay Bakshi")
    names = append(names, "Baheer Kamal")
    fmt.Println(len(names))
    fmt.Println(names)
}
```

When you run this code, you should see the following output:

```
0
2
[Tanmay Bakshi Baheer Kamal]
```

What this signifies is that when you call the make function, the array that it returns is "empty" (it has a length of zero), and you can then append to it some new elements. Once you're done, the array has a length of two, and you can see the names in the array.

However, what's different about this code snippet is that both of the append operations did not need to allocate any new memory, because the make function made it so the array already had the capacity for them internally. If you were to append more elements from here on out, it would once again do a reallocation.

At this small of a scale, this doesn't really matter, but when you start working with hundreds of thousands or millions of elements, the time saving can easily add up.

if Statements and switch Statements

Of course, an integral part of any programming language is its ability to conditionally branch out to different sections of your code. In this regard, Go supports two main ways of conditional branching:

- `if` statements
- `switch` statements

Let's start by taking a look at the classic, nearly universal "`if`" statement. In Go, these look quite similar to the `if statements` in languages such as Java, C, and Swift:

```
if <expression 1 that resolves to Boolean value> {
    <code to execute if condition 1 is true>
} else if <expression 2 that resolves to Boolean value> {
    <code to execute if condition 2 is true,
      and condition 1 is false>
} else {
    <code to execute if no previous condition was true>
}
```

The `<expression that resolves to Boolean value>` part can be a Boolean literal (`true`, `false`), a function call that returns a Boolean, a variable or constant that stores a Boolean, or an operator call that returns a Boolean (e.g., you can use `==` for checking equality, or `>` and `<` for comparison, etc.). Speaking of Boolean operators, take a look at some common Boolean and Bitwise operators in Table 2.6.

TABLE 2.6 Common Built-in Boolean and Bitwise Operators in Go

OPERATOR	WHAT IT DOES
`==`	Returns true if value on the left-hand side is equal to the value on the right-hand side.
`!`	Unary Boolean NOT operator; takes only one expression, which resolves to a Boolean, to the right. It returns true if condition to the right resolves to false and returns false if condition resolves to true.
`!=`	Returns true if value on the left-hand side is not equal to the value on the right-hand side.

(continued on next page)

TABLE 2.6 Common Built-in Boolean and Bitwise Operators in Go *(continued)*

OPERATOR	WHAT IT DOES
>	Returns true if the value on the left-hand side is greater than the value on the right-hand side.
<	Returns true if the value on the left-hand side is smaller than the value on the right-hand side.
>=	Returns true if the value on the left-hand side is greater than OR equal to the value on the right-hand side.
<=	Returns true if the value on the left-hand side is smaller than OR equal to the value on the right-hand side.
||	Returns true if EITHER: the condition on the left-hand side OR the condition on the right-hand side resolve to true. Only resolves the condition on the right-hand side if the left-hand side resolves to false.
&&	Returns true if BOTH: the condition on the left-hand side AND the condition on the right hand side resolve to true.
|	Runs a bitwise OR operation between two numbers.
&	Runs a bitwise AND operation between two numbers.

Of course, you can combine multiple Boolean operators to make a "compound condition." For example, "(if country is Canada and age is greater than 17) OR (if country is Japan and age is greater than 15)".

For example, if we want to write a short program to find out if someone's an adult, we can write the following code:

CODE LISTING 2.4 Using if statements and Boolean operators in Go

```
package main

import (
    "fmt"
)

func main() {
    age := 15
    if (age >= 18) {
        fmt.Println("Welcome!")
    } else {
        fmt.Println("You're too young.")
    }
}
```

In this program, `age` (as an expression that resolves to an integer by loading a variable) and `18` (an integer literal) are fed into either side of the `>=` operator. The operator returns a Boolean, which is then fed into the `if` statement, which determines where the code needs to jump to next.

Sometimes `if` statements:

- Make your code "ugly" because there are too many conditions to evaluate
- Always look for equality against a certain single expression

If that situation arises, you can use `switch` statements instead. Historically, `switch` statements were more optimizable than `if` statements, but compiler technology has advanced to the point where the effect of this difference is truly minimal, and only really applies in edge cases that the majority of programmers are normally unlikely to reach.

A "`switch`" enables you to evaluate a single expression and match its equality to one of many possible outcomes. For example, let's say the user enters their name, and we want to see which user of the three possible users entered their name. You can do that with the following code:

```
switch name {
case "Tanmay Bakshi":
    fmt.Println("Hi, Tanmay.")
case "Baheer Kamal":
    fmt.Println("Hi, Baheer.")
case "Michael":
    fmt.Println("Michael!")
default:
    fmt.Println("Sorry, I don't know who you are.")
}
```

Go's `switch statements` are very similar to ones in languages such as C and Swift. As in Swift, you *must* provide a `default` case, which is the case that's called when no other case's condition is satisfied. Unlike in C, you don't need to add a `break` at the end of each case, because Go doesn't pass through to other cases. It instead moves to the next line of code *after* the `switch case` block. If you don't have experience with C, no worries. You don't need it to master Go. This was simply an example to let you know how Go's control flow works with `switch` statements in comparison to the `switch` statements in other languages.

So, go ahead and implement the preceding switch statement in a main function like so:

CODE LISTING 2.5   Using switch statements to check equality in Go

```go
func main() {
    name := "Tanmay Bakshi"
    switch name {
    case "Tanmay Bakshi":
        fmt.Println("Hi, Tanmay.")
    case "Baheer Kamal":
        fmt.Println("Hi, Baheer.")
    case "Michael":
        fmt.Println("Michael!")
    default:
        fmt.Println("I don't know who you are.")
    }
}
```

When you run this code, you should see the following output:

```
Hi, Tanmay.
```

That's a primer on conditionals in Go. Using them, you can branch to different parts of your code based on whether or not a condition is fulfilled.

Loops

Loops are another integral building block that programming languages provide you with. In Go, loops are heavily simplified. In fact, just *one* kind of loop takes on the role of the *three* you usually see in other languages (for loops, while loops, do-while/repeat-while loops).

Go replaces the three loops with simply one kind of loop: the for loop. Technically, you can replicate the functionality of the other kinds of loops with a for loop as well. This means Go really just has multiple kinds of loops under the "for" banner, which represents the different kinds of loops found in other languages.

Let's start with the classic kind of loop: the "for-in" loop. This "for loop" enables you to loop through the elements in a certain sequence, such as in an array. Here's an example of using a for-in loop:

CODE LISTING 2.6 Using for-in loops in Go

```go
func main() {
    names := []string{"Tanmay Bakshi", "Baheer Kamal", "Kathy"}
    for i := range names {
        fmt.Println(i)
    }
}
```

NOTE: You're probably wondering: "I don't see the word 'in' here, so why is it called a `for-in`' loop?" The answer is simply: the "`:= range`" part of the loop is taken as "in"—for example, "`for i in names`", similar to Swift and Python.

Now, I want you to guess what this code is going to print before we actually run it. Will it print the following?

```
Tanmay Bakshi
Baheer Kamal
Kathy
```

You'd think so, right? That's what it would do with similar syntax in Swift and Java, so why not in Go? Well, this comes back to what we were talking about in Chapter 1: Go deviates from quite a few language standards. In fact, this code will actually print the following:

```
0
1
2
```

These, instead, are the indices of the elements themselves. If you'd like to get the values as well, you can use the following syntax:

CODE LISTING 2.7 Using for-in loops in Go to iterate through indices and elements

```go
func main() {
    names := []string{"Tanmay Bakshi", "Baheer Kamal", "Kathy"}
    for i, v := range names {
        fmt.Println(i)
        fmt.Println(v)
    }
}
```

Now you should see the following output:

```
0
Tanmay Bakshi
1
Baheer Kamal
2
Kathy
```

You may be wondering: Why does Go work this way? Really, from a technical and efficiency perspective, this just makes sense. For example, look at the following Swift code. I'm sure this code is simple enough to make sense without deep Swift knowledge, but we will explain it in detail to help you understand the Go syntax better.

```
let names = ["Tanmay Bakshi", "Baheer Kamal", "Kathy"]
                // Create an array of strings with 3 elements


for i in names {
    print(i)       // Print out each name in order
}
```

Of course, this code will print each name followed by a new line. However, you didn't need to create a counter to store *which* element you were on, did you?

So, if you'd like a counter, there are two ways to do it. There's the standard, generally accepted way, which uses the "enumerated" method:

```
let names = ["Tanmay Bakshi", "Baheer Kamal", "Kathy"]
                // Create an array of strings with 3 elements


for index, i in names.enumerated() {
    print(index)  // Print out the index of the name
    print(i)       // Print out each name in order
}
```

And there's the less pretty, non-standard way, which is to use a counter that you manually define and increment:

```
let names = ["Tanmay Bakshi", "Baheer Kamal", "Kathy"]
                // Create an array of strings with 3 elements
```

```
let index = 0
for i in names {
    print(index)    // Print out the index of the name
    print(i)        // Print out each name in order
    index += 1      // Increment the index
}
```

Both of these methods involve some kind of overhead. The official method involves a new call to the enumerated method in the Collection protocol, which of course, isn't terribly efficient.

Now, you're likely wondering: Isn't the second method super-efficient? *No!* It's not as efficient as you can get it to be. That's because you now have *two counters* in your code.

I know what you're saying: "But we're only incrementing one counter!" Well, here's the thing: The compiler is incrementing another counter for you. This counter is internal and is in the generated machine code. This is important because the compiler needs to know the offset from the buffer pointer to know which element we're on.

So, while the offset variable can be *hidden* from the programmer, it can't be removed altogether. Therefore, Go made the decision to just *allow the programmer to access the offset counter that it already has*. This makes it so you don't need yet another counter, which is an extra inefficiency.

However, there are some cases, quite a few in fact, where you just really don't care about the index, and you only care about the value. In those circumstances, even though Go can detect that automatically, it's still considered good practice to tell Go that you don't need it. This helps other programmers read your code more naturally and even helps the Go optimizer a bit.

Here's how you can loop through an array while ignoring the index:

CODE LISTING 2.8   Ignoring the index in a for-in loop

```
func main() {
    names := []string{"Tanmay Bakshi", "Baheer Kamal", "Kathy"}
    for _, v := range names {
        fmt.Println(v)
    }
}
```

The underscore in the `for-in` loop has the effect of telling Go: "I don't care about whatever value this resolves to, and don't let me use it." Now, if you run it, you should just see the names printed, like so:

```
Tanmay Bakshi
Baheer Kamal
Kathy
```

Now that we've discussed `for-in` loops, let's discuss an absolute classic: the C-style `for` loop!

This blast from the past consists of three main parts, which are made more flexible in Go: an expression that creates a new iterator variable, an expression that resolves to a Boolean to see if the loop needs to end, and lastly, an expression that increments the counter between loop iterations. Here's how you can use one of these loops in Go:

CODE LISTING 2.9   Using C-style for loops

```go
func main() {
    for i := 0; i < 5; i++ {
        fmt.Println(i)
    }
}
```

This code will print out 0 through 4 like this:

```
0
1
2
3
4
```

However, here's what's interesting about this `for` loop. It can emulate a `while` loop by using the following syntax:

CODE LISTING 2.10   Emulating while loops using for loops

```go
func main() {
    i := 1
    for i < 1000 {
```

```
        i += i
    }
    fmt.Println(i)
}
```

By completely neglecting to provide the first and last expressions to the `for` loop, it only ends up checking the condition you specify in the middle, and thus becomes, for all intents and purposes, a `while` loop. This code will print the following:

```
1024
```

There is technically *one* more use of a `for` loop, and that is for *channels*. Channels are a very unique feature of Go, but we need to cover a few more basics before getting there. Onward and upward!

Functions

Another great way to reuse code and make it more modular is to use functions. These are an integral part of most programming languages, and Go's functions are incredibly useful. However, at the time of writing this, they lack one KEY feature that separates clean, reusable, and less redundant code, from code that has a larger surface area for potential bugs. This is a feature the language developers are working on, however. More on this in a little bit.

You've already seen an example of a function in Go: the main function! It's written like so:

```
func main() {

}
```

If you were to run just those three lines of code, it would compile and run successfully—even though the program would do nothing. That's because this function doesn't return anything, and therefore doesn't *need* a "return" statement. We want to point this out more explicitly because some programming languages, such as C, require that the main function must return something, such as an exit code, to let the invoking process know if the program did what it was supposed to do successfully or not. This is not the case in Go.

If you'd like to create a simple function that just returns a constant value all the time, such as a 32-bit floating point representation of pi, you could write code like this:

CODE LISTING 2.11 Declaring, defining, and calling a function

```
func valueOfPi() float32 {
    return 3.14159
}


func main() {
    fmt.Println(valueOfPi())
}
```

Notice the way the function signature is structured. The return type annotation is placed directly after the name of the function and before the opening curly brace of the function block.

However, we've still got the parentheses in the way—and as you've likely guessed, these are for arguments that you may want to pass into the function.

For example, if you'd like a function that returns pi multiplied by some nonnegative integer, you could write the following function:

CODE LISTING 2.12 A new valueOfPi function that multiplies against an arbitrary constant

```
func valueOfPi(multiplier uint) float32 {
    return 3.14159 * float32(multiplier)
}
```

Once again, notice how the function signature changed. We first placed the name of the argument, then a space, and then the type of that variable.

Before we continue, there is a quick sidenote I want to make. When we ran the preceding multiplication, note how we had to manually cast the unsigned integer to a 32-bit floating point number. This is done to avoid any kind of ambiguity in the operation.

You may be wondering at this point: How does a type cast work? A type cast is able to follow a certain spec to convert bits arranged in a way that a certain type expects into a way that another type expects them to be. For example, how is a float32 represented in memory? Well, float32 in Go conforms to the IEEE 754 spec.

One bit is reserved for the sign (is this positive or negative), 8 bits are reserved for the exponent, and 23 bits are usually reserved for the fraction. However, integers are completely different: They have a very straightforward way of being represented in binary, with the value of bits increasing by a factor of 2 from right to left, with the leftmost bit usually reserved for the "sign."

So, if you want to convert an integer value to a float32 value, it's not as simple as saying "Here are the bits. Treat them as float32 type now!" Instead, you have to *tell* Go that you want to convert the bit representation from integer spec to float32 spec. You don't need to understand how this works; all you need to know is that it *does* work because the compiler and your CPU handle it for you—if the types in the cast are compatible.

Back to our code . . . Remember, we had to add the cast from uint to float32. However, in a language such as C, similar code would compile even without an explicit cast. But what would that code do? We're nearly certain only a few C developers *really* know what all those implicit operations would actually do. And, because their behavior can be undefined or change behind the scenes, you're basically asking for bugs in your code.

While Go code can sometimes be quite terse because of this rule, it's worth it for the safety, which as we discussed, is something that Go values highly.

Back to the function . . . In order to call it, we can do this:

```
valueOfPi(2)
```

An example of using this expression in a real program is:

CODE LISTING 2.13   Calling the new valueOfPi function with a multiplier

```
func main() {
    fmt.Println(valueOfPi(2)) // Prints tau (2 * the value of Pi)
}
```

Multiple arguments are delimited by commas at both the function declaration and the call site. Here's a similar function that not only takes the non-negative multiplier to multiply by, but also a signed integer that we want to subtract from pi.

CODE LISTING 2.14   Implementing operateOnPi to offset and multiply pi by constants

```
func operateOnPi(multiplier uint, offset int) float32 {
    return (3.14159 — float32(offset)) * float32(multiplier)
}
```

Notice how the function signature kept the same theme—the arguments are listed, comma delimited, within the parentheses, and the name of each argument and its type annotation are listed in that order, delimited by a space.

When you want to call that function, simply delimit the two arguments you pass with a comma, like so:

CODE LISTING 2.15   Calling the new operateOnPi function

```go
func main() {
    fmt.Println(operateOnPi(2, 1))   // Prints the value that
                                     // results from the
                                     // expression:
                                     // (3.14159 - 1) * 2
}
```

Let's tie this back to what we learned a little while ago: variables. Remember, when you are initializing a variable, you're really just writing out the following:

```
<name> := [expression that resolves to the value you'd like
    ↪ to store]
```

The expression on the right may be a call to a function that returns what you want to store, it may be another variable, it could be a literal, or just *anything* that resolves to the value you expect.

There is another way, but it still follows the same theme:

```
var <name> <type>
<name> = [expression that resolves to the above <type>]
```

So, if you'd like to call the function we built and store the value in a variable, the normal way to do it is:

```
tau := operateOnPi(2, 0)
```

Now this variable contains the value of Tau. However, we can *still do more.* What if you want to return *multiple* values? Say we're building a function that returns both a name AND an age, how would we handle that?

Here's an example that does just that:

```go
func nameAndAge(uid int) (string, int) {
    switch uid {
```

```
    case 0:
        return "Baheer Kamal", 24
    case 1:
        return "Tanmay Bakshi", 16
    default:
        return "", -1
    }
}
```

This function takes a user ID, matches it using a `switch` statement, and returns the name and age associated with that user ID. Once again, the function signature is still familiar, and the change makes sense. Instead of a standalone type annotation between the function name with its arguments and the code block, there are multiple type annotations, delimited by commas, and enclosed in parentheses.

At the site of the function return, we *don't use the parentheses*, instead we just delimit the different expressions that resolve to the return values with commas. In this case, all of those expressions are literals, but they don't have to be.

However, using this function may not be as intuitive as you expect. Depending on which languages you've coded in before, it may look like Go is returning a type such as a "tuple," but it isn't really. It's actually returning two separate, distinct values. This means that the following code will fail to compile:

```
user := nameAndAge(0)
```

The reason why it fails is because Go is expecting the expression on the right to resolve to *two* values, while there's only *one* on the left side of the `:=` operator. So, as the function signature demands, in order to fix this issue, you need to have two variables on the left side, delimited by a comma:

```
userName, userAge := nameAndAge(0)
```

Something to keep in mind here is that, as long as there is at least *one* new variable on the left side of the `:=` operator, you must use, well, the `:=` operator. If there are no new variables, then you must use the simple `=` assignment operator. For example, this code makes sense:

```
var userName string
userName, userAge := nameAndAge(0)
```

This code works because `userAge` is still a new variable, so Go tolerates the pre-declared userName variable on the left. However, in the following code, both are already declared.

```
var userName string
var userAge int
userName, userAge = nameAndAge(0)
```

In the preceding snippet, you're not allowed to use the `:=` operator, you must use the = operator.

However, what if you don't care about, say, the `name`, and you only care about the `age`? To understand how you can ignore a return value, let's look at this sample program:

CODE LISTING 2.16   Returning multiple values from a function

```
func nameAndAge(uid int) (string, int) {
    switch uid {
    case 0:
        return "Baheer Kamal", 24
    case 1:
        return "Tanmay Bakshi", 16
    default:
        return "", -1
    }
}


func main() {
    userName, userAge := nameAndAge(0)
    fmt.Println("User age:")
    fmt.Println(userAge)
}
```

This looks like perfectly valid syntax, so let's compile it! And . . .

Compile error . . .

Why, you may ask? It's because we declared userName, but we never used it. Most programming languages would ignore it, some would throw a warning. However, Go completely stops the compilation process, and says you *must* fix this code before continuing. This is what we meant in Chapter 1 by "Go has a very strict compiler"!

The reason why Go throws an error is because when you don't use a variable but you do go through the effort of naming and declaring it, it's usually indicative of a bug. For example, what if you forgot to write the logic that deals with that value? Usually, though, it's just a value that you don't need. If so, then to tell the compiler you'd like to ignore it, you can assign it to an underscore:

```go
func main() {
    _, userAge := nameAndAge(0)
    fmt.Println("User age:")
    fmt.Println(userAge)
}
```

Now, this code will compile, and print:

```
24
```

One more thing to note: The underscore is not considered a new variable, but userAge is, and therefore we use the := operator. If we had already declared userAge, then because underscore is not a "new" declaration, we'd have to use the = operator, like so:

```go
func main() {
    var userAge int
    _, userAge = nameAndAge(0)
    fmt.Println("User age:")
    fmt.Println(userAge)
}
```

Another way you can use functions in Go is, of course, to pass them to other functions! For example, what if you need a function that calls another function, but you don't know at compile time *which* function it needs to call? You could write code like this:

CODE LISTING 2.17   Passing function pointers to other functions

```go
func runMathOp(a int, b int, op func(int, int)int) int {
    return op(a, b)
}
```

```go
func add(a int, b int) int { return a + b }
func sub(a int, b int) int { return a - b }
func mul(a int, b int) int { return a * b }
func div(a int, b int) int { return a / b }

func main() {
    a, b := 9, 6
    fmt.Println(runMathOp(a, b, add))
    fmt.Println(runMathOp(a, b, sub))
    fmt.Println(runMathOp(a, b, mul))
    fmt.Println(runMathOp(a, b, div))
}
```

Notice the function signature of the runMathOp function. It is taking an argument called "op," the type of which looks like a regular function signature, just without the function name and argument names—only the argument types and return type(s) remain.

When you call the runMathOp function, you can simply pass to it another function just as you would any other parameter, such as add or mul, as long as it matches the function signature specified in the runMathOp declaration.

One last very powerful feature you should be aware of before we continue is called *defer*. With defer, you tell Go to execute some code just before your function is about to return. This code will be executed after the return expression has been evaluated, but right before the actual return to the caller occurs. The syntax is very simple: You simply write the defer keyword and run a function call right after it. So, for example, look at the following code:

```go
package main

import (
    "fmt"
)

func test(x int) int {
    defer fmt.Println(x)
    y := x + 1
    fmt.Println(y)
    return y
```

```
}

func main() {
    test(5)
}
```

`println(x)` is a function call, and that call is being passed to the defer key-word within the `test` function. So, that print won't be executed until just before the test function returns. Because of this, you should see the following output:

```
6
5
```

You can also define a new function *inline* with the defer and call it inline, like so:

CODE LISTING 2.18   Defining and calling a function inline for defer

```
package main

import (
    "fmt"
)

func test(x int) int {
    defer func() {
        fmt.Println("this is being called from an inline
            ↪ function")
        fmt.Println("I can put multiple expressions inside
            ↪ of here!")
        z := x - 1
        fmt.Println(z)
    }()
    y := x + 1
    fmt.Println(y)
    return y
}

func main() {
```

```
    test(5)
}
```

Because you're not just *defining* a function inline, but also *calling* it on line 9 with the two parentheses, which is what defer requires, this is valid syntax. You should see the following output when you run this code:

```
6
this is being called from an inline function
I can put multiple expressions inside of here!
4
```

Now that we've covered the basics surrounding functions, let's talk about what you *cannot* do with Go: *generics.*

Generics are one of the most beloved features of the languages that support them. At least, most of them (we're looking at you, Java and C++!).

In Swift and Julia, for example, generics enable you to write less code that supports more types and has more readability and flexibility. A language such as Python doesn't really count for this comparison because you're duck typing in it anyway! It doesn't care about the types of your variables in the first place.

However, generics are being worked on by the Go team. Here's a sneak peak of some code you will be able to compile in a future version of the Go language:

```
func Print[type T](s []T) {
    for _, v := range s {
        fmt.Print(v)
    }
}
```

This code may look complex, but it's really quite simple. Previously, let's say you wanted to write a function that would print every element of an array of integers. You'd write something like this:

```
func Print(s []int) {
    for _, v := range s {
        fmt.Print(v)
    }
}
```

And it works! But then if you want to do the same thing with strings, you have to do this:

```
func Print(s []string) {
    for _, v := range s {
        fmt.Print(v)
    }
}
```

And you have to do something similar for floats, unsigned integers, different sizes of integers and floats, and so on. It's really not fun and results in a lot of repeated code. However, let's take a look at a generic version of the function:

```
func Print[type T](s []T) {
    for _, v := range s {
        fmt.Print(v)
    }
}
```

You can now pass in *any* kind of array, and it'll print the elements! And, because all the calls and operations are disambiguated at compile time, there's no runtime overhead to this kind of polymorphism. However, remember that Go wasn't designed for and around generics from the ground up, so some functionality may be a bit of a "band-aid" solution to work around language design shortcomings that inhibit truly powerful generics.

For example, in Go, if you wanted to build a generic function that takes two variables of the same type and returns the smaller one, you could write:

```
type numeric interface {
    type int, int8, int16, int32, int64, uint, uint8, uint16,
        ↪ uint32, uint64, float32, float64
}
```

```
func min(type T numeric)(a T, b T) T {
    if a < b {
        return a
    }
    return b
}
```

Whereas, in a language such as Swift, built around generics from the core, you could do this:

```
func min<T: Comparable>(a: T, b: T) -> T {
    a < b ? a : b
}
```

Which basically translates to: "Assume there's a type called T, and T must conform to the 'Comparable' protocol. This means we know there is a less than and greater than operator implemented for this type. This function takes two arguments, a and b, both of this type T, and we also return a value of type T. Then, check if a is less than b. If it is, return a; otherwise, return b."

By default, lots of different types *already* conform to the Comparable protocol, and even your own classes and structures can add their own conformance to it, too.

So, while the Go code is a bit inelegant, it gets the job done, and it gets it done quite well, too.

Structures

Go is *not* an object-oriented programming language. It has no concept of a "class" or an "object". Everything is a value. Even references are values, meaning that they're "pointers" represented as values.

So, to aid the programmer in storing multiple pieces of related data, Go has the concept of structures. Structures enable you to store some data as a contiguous block of memory. Let's go back to our user name and age example. We could represent that as a structure:

```
type User struct {
    Name string
    Age int
}
```

Now, the size of User in bytes is the size of string in bytes + the size of int in bytes. Quite literally, the bytes for these variables are just laid out sequentially, one after the other, in memory.

If you'd like to create a new instance of this structure, you can use the following syntax:

```
myUser := User{"Kathy", 18}
```

Now, if you want to print out the different pieces of data stored within the structure, you can simply use the "`.`" operator:

```
fmt.Println(myUser.Name)
fmt.Println(myUser.Age)
```

Now, let's take a look at an example of using a structure in a program. To do so, let's adapt our previous example from the functions section:

CODE LISTING 2.19   Using a structure to replace multiple return values in a function

```
type User struct {
    Name string
    Age int
}

func nameAndAge(uid int) User {
    switch uid {
    case 0:
        return User{"Baheer Kamal", 24}
    case 1:
        return User{"Tanmay Bakshi", 16}
    default:
        return User{"", -1}
    }
}

func main() {
    user := nameAndAge(1)
    fmt.Println("User age:")
    fmt.Println(user.Age)
}
```

As you can see, we can use our new `User` structure type just like any other type in this code. Remember, though, that structures are *value types*, not *reference types*, like classes are. Therefore, something like this may give you unexpected results, depending on what you've programmed in before:

CODE LISTING 2.20   Mutating a structure passed to a function as an argument

```go
func incrementAge(user User) {
    user.Age++
    fmt.Println(user.Age)
}


func main() {
    kathy := User{"Kathy", 19}
    incrementAge(kathy)
    fmt.Println(kathy.Age)
}
```

This would print:

```
20
19
```

This is because the function that's responsible for incrementing the age variable is simply taking a User type, which means that Go will make a copy of the structure, and pass the function to that copy. The function then modifies that copy, prints the age, it goes out of scope, and the copy is deleted. However, back in the main function, the original "kathy" has never been touched, so it prints the original age.

If instead you want to have a function that modifies a structure or other value in place, you must pass a reference to it. That can be done with the following syntax, which should be familiar to you if you've programmed in C:

CODE LISTING 2.21   Mutating a structure through its pointer passed to a function

```go
func incrementAge(user *User) {
    user.Age++
    fmt.Println(user.Age)
}


func main() {
    kathy := User{"Kathy", 19}
    incrementAge(&kathy)
    fmt.Println(kathy.Age)
}
```

Note that there are only two changes in this code. The type annotation in the `incrementAge` function has changed to `*User` instead of `User`, meaning it's a reference. Plus, at the call site in the `main` function, we're no longer passing `kathy`, we're feeding `kathy` into the "`&`" operator, which fetches a reference to `kathy`, and then we pass that reference into the function.

When you run this code, you should see:

```
20
20
```

Bingo! Now, we would like to point out here that this is NOT a pointer, despite what your C instincts may be telling you! This is a *reference* that's handled by Go. So, within the function, you don't need to dereference the pointer before accessing the `age` property. Remember, in Go, we did:

```
fmt.Println(user.Age)
```

Whereas, in C, with a pointer, we'd need to do:

```
printf("%d\n", user->age);
```

Or:

```
printf("%d\n", *user.age);
```

As you may have noticed, in C we wouldn't be able to use the "`.`" operator unless we dereferenced the pointer ourselves, or we would need to use the "`->`" operator to dereference it for us automatically. Luckily, in Go, we have no such restriction, because this isn't really a pointer in the first place; it's a reference.

Another thing you can do with structures in Go is assign functions to them! For example, what if we need a simple function for our `User` structure that could return a pretty string like "Baheer Kamal is 24 years old!"? We could write a function like this:

```
func (user User) prettyString() string {
    return fmt.Sprintf("%s is %d years old!", user.Name,
        ↳ user.Age)
}
```

As you can see, the function signature has changed in one key way: After the "`func`" keyword, we have a new argument name and type annotation in parentheses before the rest of the usual function signature. This tells Go "make this an instance method on the `User` `structure`, and pass that instance into the function with the name 'user'."

Now, you can call this function like so:

```
func main() {
    kathy := User{"Kathy", 19}
    fmt.Println(kathy.prettyString())
}
```

This will print:

```
Kathy is 19 years old!
```

Great! Now, for some extra practice, let's port over the `incrementAge` function to an instance method:

CODE LISTING 2.22   Mutating a structure using a value receiver

```
func (user User) incrementAge() {
    user.Age++
    fmt.Println(user.Age)
}


func main() {
    kathy := User{"Kathy", 19}
    kathy.incrementAge()
    fmt.Println(kathy.Age)
}
```

All right, looks good. If you run it now, you should see:

```
20
19
```

Hold it. Why didn't the instance method update the structure? Well, look again at the function signature, and notice the type annotation is User, not *User. That's right, even these functions are passed a copy of the structure. Let's fix that by using a "pointer receiver":

CODE LISTING 2.23   Mutating a structure using a pointer receiver

```go
func (user *User) incrementAge() {
    user.Age++
    fmt.Println(user.Age)
}

func main() {
    kathy := User{"Kathy", 19}
    kathy.incrementAge()
    fmt.Println(kathy.Age)
}
```

This time, you can see we did *not* need to update the call site, and instead only updated the function signature saying that we want a reference to the User on which we are calling this function.

Now you should see the following output:

```
20
20
```

Great! You've now learned the basics of structures in Go. Of course, there still is quite a bit more to learn, like Reflection, but we'll get into that in a later chapter.

Interfaces

In Go, we also have what is known as "interfaces," which are similar to the interfaces in Java or protocols in Swift. However, they do have some more limitations (again, Go can be quite "classic" in this regard).

An interface enables you to define a group of functions that a certain structure needs to have, without tying them to an implementation. For example, you could say that "to be compatible with the 'Living' interface, you must have an incrementAge function that takes no arguments and returns nothing."

Let's take a look at a simple group of structures without an interface:

CODE LISTING 2.24   Similar structures without a backing interface

```go
type Person struct {
    Name string
    Age int
}

type Dog struct {
    Name string
    Owner *Person
    Age int
}

func (person *Person) incrementAge() {
    person.Age++
}
func (person *Person) getAge() int {
    return person.Age
}

func (dog *Dog) incrementAge() {
    dog.Age++
}
func (dog *Dog) getAge() int {
    return dog.Age
}
```

NOTE: The eagle-eyed among you may have noticed a "mistake" here, at least according to what we've talked about before. In the getAge() function, we have a "pointer receiver"—i.e., the type annotation for the structure we're attaching to is *Dog not Dog. But we're not changing anything, so why does it need to be a pointer? We'll get to that in a bit.

Now, what if we wanted to write a function that would be able to call incrementAge and then print out the new age, regardless of what *kind* of living being it is? What if we could abstract away the species? We don't need full-blown parametric polymorphism (`generics`), we just need `interfaces`. For example, we can define the following `interface`:

CODE LISTING 2.25   An interface to standardize the two structures

```
type Living interface {
    incrementAge()
    getAge() int
}
```

What we've done is we've told Go "if a structure implements these two functions, let me just refer to it as Living, and assume it only has the functionality defined in the `Living interface`." Therefore, we can write a function like this:

CODE LISTING 2.26   A function that mutates an instance of a structure conforming to the interface

```
func incrementAndPrintAge(being Living) {
    being.incrementAge()
    fmt.Println(being.getAge())
}
```

And then call it like this:

CODE LISTING 2.27   Calling incrementAndPrintAge with two structure instances

```
func main() {
    harley := Person{"Harley", 21}
    snowy := Dog{"Snowy", &harley, 6}
    incrementAndPrintAge(&harley)
    incrementAndPrintAge(&snowy)
}
```

Now, don't think too hard about this code. We'll walk through it, and through the Go compiler's thought process, in more detail in just a moment. Right now, just run it, and admire the output of your hard work:

```
22
7
```

Hooray! But how did that happen? Let's dig a bit deeper.

First, let's look at the `incrementAge` and `getAge` functions. Notice that, for both of them, they really are receiving pointer values, not regular values. This means that `Person` and `Dog` both do *not* have the `incrementAge` and `getAge` functions—instead, `*Person` and `*Dog`, the *reference types*, implement those functions.

So, let's take a look again at our `interface`:

```
type Living interface {
    incrementAge()
    getAge() int
}
```

The `interface` says "any type that implements these two functions can also be referred to as the `Living` type." And, what do you know, `Person` and `Dog` don't implement them, instead `*Person` and `*Dog` do. Therefore, when we call the `incrementAndPrintAge` function, if we just pass the variables directly as:

```
incrementAndPrintAge(harley)
incrementAndPrintAge(snowy)
```

. . . we'd get an error because the types of those variables don't implement the functions needed to conform to the `Living` interface. However, when we pass the reference:

```
incrementAndPrintAge(&harley)
incrementAndPrintAge(&snowy)
```

. . . then we're not passing `Person` and `Dog`, we're passing `*Person` and `*Dog`, which do conform properly to the `interface`.

So, this has proven something to us, but it's basically subliminal: *The functions in an interface can either all be pointer receivers, or all be value receivers. There is no in-between.*

This is definitely not optimal. But you may ask, why exactly is this the case? Well, let's make our code a bit better, and in the `getAge()` functions, let's replace the reference types with value types:

```
func (person Person) getAge() int {
    return person.Age
}
```

```
func (dog Dog) getAge() int {
    return dog.Age
}
```

Now, let's see if our types conform properly.

`Dog` and `Person` unfortunately still don't conform to `Living`, because they only implement `getAge` and not `incrementAge`. However `*Dog` and `*Person` now *also* don't conform, because they only implement `incrementAge` and not `getAge`. Now you can see why this is a problem. No type in this code currently conforms to `Living`!

Therefore, our interface becomes essentially useless.

There are definitely better ways Go could handle these situations, but for now, if even one function in an `interface` requires a pointer receiver, *all* your functions will need a pointer receiver.

Errors

Go has a somewhat unique way of handling errors. In most popular languages, a function can "throw" an error and the caller can "catch" the error. This approach is still possible in Go, but it's uncommon and not encouraged. Rather, in Go you're encouraged to return a type called `error`.

`error` is an interface that only expects the underlying type to implement a single function. This function is called `Error`, takes no arguments, and returns the error it represents as a human-readable string. The definition for the `error` interface is as follows:

```
type error interface {
    Error() string
}
```

You can create your own error types very easily! The following is a simple implementation of an error:

```
type errorString struct {
    s string
}

func (e *errorString) Error() string {
    return e.s
}
```

The `errorString` structure now conforms to the `error` interface, and a function that you define could return this as an `error`. However, you may ask, what's the point of an interface in the first place? Why not just make `error` a structure instead, and that way, your function can just store the `error` as a string directly?

Well, the answer is simple: Errors are complex. Let's say you get an error from a function that's meant to make a request to an HTTP endpoint. The error could occur at multiple points, and you may have not been able to *send* the request because of a firewall or lack of an internet connection. Maybe the response code was 500 (Internal Server Error), so it was the server's fault!

So, instead of simplifying what counts as an `error` to just a string, Go makes it so you can define your own `error` structures to contain custom error information. Then, because you're forced to implement the `Error()` function to return a human-readable string, you can still print out any runtime errors to debug them. These strings can be assembled based off of the other content of the structure—such as, in the HTTP example, a response code.

This unorthodox approach to error handling leads to somewhat inelegant code, but usually reduces the chance of a crash due to an unhandled error. To see how, let's go over an example of using `error` in a function.

When it comes to defining new kinds of simple errors, the standard in Go is to define them as global variables. This allows code from outside the package to see, check, and deal with the error that was returned more effectively.

In the following example, we use the `errors` package to help us create the `DivisionByZero` error. We then define a `Divide` function, and if this function is asked to divide a number by zero, it returns our custom error.

Right now, you don't need to worry about the second line of code that deals with importing the errors module. You'll learn more about modules and how they work in the next chapter. For now, all you need to know is that it affords us the functionality to define our own simple errors.

CODE LISTING 2.28 A simple DivisionByZero error

```
package main

import (
    "errors"
    "fmt"
```

```
)

var DivisionByZero = errors.New("division by zero")

func Divide(number, d float32) (float32, error) {
    if d == 0 {
        return 0, DivisionByZero
    }
    return number / d, nil
}

func main() {
    n1, e1 := Divide(1, 1)
    fmt.Println(n1)
    if e1 != nil {
        fmt.Println(e1.Error())
    }
    n2, e2 := Divide(1, 0)
    fmt.Println(n2)
    if e2 != nil {
        fmt.Println(e2.Error())
    }
}
```

You should see the following output:

```
+1.000000e+000
+0.000000e+000
division by zero
```

As you can see, the first division goes through properly, but the second one returns an error!

Now, there are still times where the program crashes due to something unexpected happening, and these types of error are usually handled with a try/catch block. For any crash that happens in the block, the program will catch it and try to handle the error. So, while Go supports this feature, it's not implemented the same way it is in other languages, such as Java and C++.

For any unexpected error, Go will throw a *panic*, which kills the program. In fact, some packages can also throw a panic if they encounter unexpected behavior. Alongside panics, Go has another concept called *recover*. This function will catch any panic that either a package or the language throws.

The dance between the function stack, panic, and recover goes something like this: The function that throws a panic will immediately return but will run its defer operation beforehand if there is one. Then, the caller function will *also* immediately return to its caller, and so on until you hit the root of the call stack (which is the program invocation). Each return will execute the defer for its corresponding function if there is one. In any of the defer blocks, you are allowed to stop the loop of function returns by calling the recover function and doing something with the error that came from the top of the call stack. If, however, you reach the root of the call stack and you haven't run a recover, the program will crash, and Go will print out the error for you.

To understand this a bit better, take a look at the following code:

CODE LISTING 2.29   The behavior of panic: single function call

```go
package main

import (
    "errors"
    "fmt"
)

var SampleError = errors.New("This is a test error")

func testRecover() {
    defer func() {
        if recover() != nil {
            fmt.Println("got an error!")
        } else {
            fmt.Println("no error")
        }
    }()
    panic(SampleError)
    fmt.Println("Hello!")
}
```

```
func main() {
    testRecover()
}
```

In the `testRecover` function, the first thing we do is define a defer function that simply runs the recover function, and checks if the return value is not nil. If so, then we know we got a panic, and we can print that out. Otherwise, there was no error, and this defer is just being called before a normal function return. Because we have called the recover function, the panic was handled, and the panic will not propagate to the caller of `testRecover`, which is `main`.

If you run the code as is, you should see:

```
got an error!
```

If you were to comment out the panic, you'd notice the code prints:

```
Hello!
no error
```

Now, let's say you refactor the code to the following:

CODE LISTING 2.30   The behavior of panic: two function calls deep

```
package main

import (
    "errors"
    "fmt"
)

var SampleError = errors.New("This is a test error")

func testPanic() {
    panic(SampleError)
    fmt.Println("Hello from testPanic!")
}

func testRecover() {
    defer func() {
```

```
        if recover() != nil {
            fmt.Println("got an error!")
        } else {
            fmt.Println("no error")
        }
    }()
    testPanic()
    fmt.Println("Hello from testRecover!")
}

func main() {
    testRecover()
}
```

In this case, the panic is not being handled within `testPanic`, so it propagates to `testRecover`, which is forced to immediately return without printing. It does, however, run its defer, which runs recover, which prevents the panic from propagating to `main`, which happens to be the root of the call stack. This means that the program will not crash.

All in all, after the panic & recover dance, the output of the program is the following:

```
got an error!
```

With that, you've learned about all the basic syntax that will help you on your Go programming journey. Next, in Chapter 3, let's learn some more advanced concepts by building some applications, such as a pathfinder and an implementation of the Game of Life!

Exercises

1. How does Golang make sure that values of constants do not change accidentally or intentionally?

2. a. Give one example where Go compromises with a language feature to improve the speed of compilation.

    b. Give three instances where Go deviates from other programming language standards for the purpose of efficiency.

3. How are variables and references to variables treated when the scope in which they exist is over?

4. In the following code you printed the value of i and found it was 1024. What will happen if the line that prints the value of i and the curly brace above it are swapped? Predict the answer, and then run it to test your prediction.

    Additionally, modify this program to write 1, 2, 4, . . . , 512, 1024 in a single column without commas between the numbers.

```
func main() {
    i := 1
    for i < 1000 {
        i += i
    }
    fmt.Println(i)
}
```

5. What is code reuse? Name the concepts used in this chapter that help you reuse your code.

6. How can a function directly mutate one of its input arguments?

7. How is a structure in Go different from structures in Swift and from classes in Swift and Java?

8. Write a simple program that takes a number from the user and panics if the number is not prime, but allow the panic to be recovered if the number is even, and in that case, tell the user that even numbers are not prime, excluding 2, as it is prime.

Go Modules

Welcome to Chapter 3! Now that you're aware of the basic building blocks of the Go language, you may think it's time to start implementing more complex apps! However, before we get to that, it's important to learn how to work together with the larger community of programmers. To do this, you must learn how to consume and create *packages*, which enable code to be distributed at scale.

Once you've finished this chapter, you'll be able to answer the following questions:

- What are packages and why are they important for a programming language's ecosystem?
- How do package managers work, and what is their role in developing large applications?
- How are Go modules structured?
- How can you use built-in and third-party modules in your apps?
- How can you build your own Go modules?

Most programming languages implement "package management" systems to enable this distribution of code. When done right, package management systems enable programmers to be more efficient and cooperative. Take, for example, Python's PIP, Swift's Swift Package Manager, Ruby's RubyGems, and of course, Go's "Go modules."

Plus, when an amazing ecosystem of packages, both built-in and community supported, are accompanied by incredible language features, such as Go's unique concurrency and fast compilation, programming becomes easy and fun.

Programming languages are really defined by the "packages" of functionality they provide. Take Python, for instance. Python has an extensive set of usually well-optimized built-in packages that extend its "standard library." For example, "json," "urllib," "pickle," "os," "sys," "sqlite," and more. These packages are one of the pillars of Python's success, even despite its shortcomings from a technical perspective—you only need to install the language, and you can make and receive HTTP requests, parse JSON and CSV files, serialize data to disk, control OS features, and even use SQLite databases.

One more key to its success is that it not only provides basic functionality with built-in packages, but also offers an easy way for others to distribute their own packages and leverage others' packages. If a system like this isn't easy to use, centralized, and a feature that's built alongside the language itself, you can end up with a fragmented, messy, bug-spawning ecosystem such as that of Java or older versions of Swift for platforms other than iOS.

Using Built-in Packages

The first example we'll explore is a relatively simple application that can grab data from the "OMDb" API. OMDb stands for "The Open Movie Database," and as the name suggests, they provide programmatically accessible information about a vast array of movies. Think "IMDb, but cheaply API callable."

Specifically, let's build a small application that lets us search for movies by their name. We'll start the application with the usual, plus a few extras—don't worry, we'll explain what they mean in just a moment!

```
package main

import (
    "encoding/json"
```

```
    "errors"
    "io/ioutil"
    "net/http"
    "net/url"
    "strings"
)
```

Of course, the first line of code just tells Go which package we're writing code for. Then, we come to the imports. So far, you've only had to use the "fmt" library, but this time we're using quite a few other libraries, as you can see in Table 3.1.

TABLE 3.1   The Purpose for Importing Each Package in the OMDb Searcher App

PACKAGE	PURPOSE
encoding/json	Used to encode/decode (and marshal/unmarshal) JSON objects
errors	Used to raise errors
io/ioutil	Input/Output utilities (this will be used for reading the data stream from the REST API)
net/http	Implementation of HTTP Client and Server (we will only use the client features)
net/url	Implementation of utilities to deal with URLs (including HTTP URLs)
strings	Implementation of string utilities

These packages will provide the building blocks upon which we can implement our own logic.

NOTE: Something to keep in mind is that, in Go, every import *must be used*. Otherwise, unlike some other programming languages, which would either continue silently or throw a warning, Go will throw an error and refuse to compile your program. This is for both speed and safety—an unused import may point to the presence of a bug. For instance, you may have forgotten to implement some important functionality.

After the imports, we're going to create a new constant called "APIKEY," which stores the OMDb API key. An API key is used to tell an API who you are, so that the service can implement features such as rate limiting, permissions, and billing.

```go
// omdbapi.com API key
const APIKEY = "193ef3a"
```

Before we continue, we want to make a quick note about the API: We're going to implement two functions that call two of the many API endpoints provided to us. Both of these functions will implement movie search functionality. One will implement searching by title, and the other will implement searching by Movie ID (the unique identifier of a movie for IMDB).

When you call either of these endpoints, you get a JSON response that contains an object that conforms to a certain "spec" (specification). What we'll do now is implement that same spec in a structure in Go. This way, when we get the JSON string as a response, we can just tell Go to marshal that JSON into the structure!

Here's the structure we'll build:

```go
// The structure of the returned JSON from omdbapi.com
// To keep this example short, some of the values are not
// mapped into the structure
type MovieInfo struct {
    Title       string `json:"Title"`
    Year        string `json:"Year"`
    Rated       string `json:"Rated"`
    Released    string `json:"Released"`
    Runtime     string `json:"Runtime"`
    Genre       string `json:"Genre"`
    Writer      string `json:"Writer"`
    Actors      string `json:"Actors"`
    Plot        string `json:"Plot"`
    Language    string `json:"Language"`
    Country     string `json:"Country"`
    Awards      string `json:"Awards"`
    Poster      string `json:"Poster"`
    ImdbRating  string `json:"imdbRating"`
    ImdbID      string `json:"imdbID"`
}
```

There are quite a few other pieces of data that we get from the API, but we've omitted them from the structure for the sake of brevity. There is also some extra syntax in this structure that we haven't introduced you to yet—it's the part that comes after the type annotation, enclosed in backticks. For example:

```
Actors      string `json:"Actors"`
```

What exactly does the `json:"Actors"` part do? This tells the JSON package that, when you need to marshal a JSON string into this structure, the "Actors" key's value should be placed into this variable (the "Actors" string). This allows you to have a name for the key in JSON that is different from the variable name that stores it in the structure.

To make this even clearer for you, in Chapter 2 we introduced a simplified version of structures. We explained you need to provide a variable name and type annotation within the structure, but there's another piece of variable information that Go can accept, albeit optionally. This is called a "tag." The tag contains extra info about the variable and what different parts of your code need to know about it.

In this case, we're using the tag, which contains the JSON key, to support JSON marshaling.

Now that we have a structure containing information we get from the API, let's implement the calls to the API. Before that though, we must write the function that's actually responsible for sending the HTTP GET requests to the API. Here's a simple function that implements this logic:

```
func sendGetRequest(url string) (string, error) {
    resp, err := http.Get(url)
    if err != nil {
        return "", err
    }

    defer resp.Body.Close()
    body, err := ioutil.ReadAll(resp.Body)
    if err != nil {
        return "", err
    }

    if resp.StatusCode != 200 {
        return string(body), errors.New(resp.Status)
    }
    return string(body), nil
}
```

The way this function works is simple, even though it may look complex because it's the first large function you deal with in Go. Let's start by deciphering the function signature:

```go
func sendGetRequest(url string) (string, error)
```

This signature is quite straightforward. The function takes only a single argument, called url, of type string. However, it returns *two* separate values: a string that represents the response of the request as a string; and an error type, which should be nil but can contain a value if an error occurred somewhere in making the request or parsing the response.

After the signature comes the main body of the function. Because this is the first real function you've been exposed to, let's go through it section by section in Table 3.2.

TABLE 3.2   The Functionality of Each Section of Code Within sendGetRequest

CODE	WHAT IT DOES
```go resp, err := http.Get(url) if err != nil { return "", err } ```	Run the actual GET request using the Get function from the http module, and pass it the url passed into this function as an argument. Then, store the response in resp and err, and make sure err is nil. If it isn't, return from this function early with an empty response and this error.
```go defer resp.Body.Close() body, err := ioutil.ReadAll(resp.Body) if err != nil { return "", err } ```	Tell Go that "Before this function returns to its caller in the future, make sure to close the Body input stream from the response." Then, we use the io package to read the bytes we get from that response Body. This can return an error, so we run logic similar to that in section 1—if there is an error, return early with an empty response and the error we got.
```go if resp.StatusCode != 200 { return string(body), errors.New(resp.Status) } ```	Check the status code we got, and if it's not 200 (meaning everything's fine), then return the body of the response (which may contain useful information) and create a new error, the description of which is the entire status of the response.
```go return string(body), nil ```	This final section is where we hope to reach—this means there were, to our knowledge, no errors in the pipeline, and that we can return the body as a string, with no error.

Wonderful! With this function completed, we now have the ability to send GET requests and handle some common errors that may come up as we try to do so. Now, let's implement searching.

Remember, we need to implement searching two ways: by title, and by movie ID. Let's start with searching by title or name:

```go
func SearchByName(name string) (*MovieInfo, error) {
    parms := url.Values{}
    parms.Set("apikey", APIKEY)
    parms.Set("t", name)
    siteURL := "http://www.omdbapi.com/?" + parms.Encode()
    body, err := sendGetRequest(siteURL)
    if err != nil {
        return nil, errors.New(err.Error() + "\nBody:" + body)
    }
    mi := &MovieInfo{}
    return mi, json.Unmarshal([]byte(body), mi)
}
```

The function looks a lot simpler! Let's again start by looking at the function signature:

```go
func SearchByName(name string) (*MovieInfo, error)
```

The function signature looks straightforward at first glance: You're simply creating a function that takes a single argument and returns two values. However, upon closer inspection, you realize that the first value being returned, the MovieInfo struct that will be unmarshaled from the JSON response, is actually a pointer!

Why is this? It's because we also have another return value: error. When there's an error, it means we do *not* have a MovieInfo structure to return . . . So, what do we return in its place? We could return a blank structure with a bunch of placeholder values, but that's inelegant. You may ask, why not just return nil? Well, if the type annotation for the function were just MovieInfo, as a value and not a pointer, then Go would not allow us to return nil, because there wouldn't be a way to represent that in memory.

Instead, when we declare this to be a pointer, Go gives us permission to return nil, because the pointer is allowed to point to nothing. So, if we need to return an error, we can return a nil value with an error, or a value with a nil error if there were no issues.

In terms of the logic itself, the function is essentially just following these steps:

1. Build a set of URL parameters that contain the API key and the name of the movie we want to search.
2. Compose the REST API URL we want to query, along with the parameters from step 1.
3. Make the request to the site, and if there is an error, return a nil pointer and the error.
4. Create a new MovieInfo value, get a pointer to it, return the pointer, and unmarshal the response string into the value through the pointer. If there is an error from unmarshaling, return that error as well.

Of all the steps, step 4 is likely the only one that's complex, because it ties into the last two lines of code in the function. Let's take a look at those in more detail:

```
mi := &MovieInfo{}
return mi, json.Unmarshal([]byte(body), mi)
```

The first line in this snippet is simple: We create the structure, get a pointer with the ampersand operator, and store that address in the mi variable. The second line is where things get a bit more complex. If you were to break it down, you'd realize the first value that needs to be returned is an expression that's easily resolvable: It's just a value in a variable.

However, the second value that needs to be returned is not so easy to resolve. You need to call a function (which in this case is Unmarshal), get the return value, and then use that as a return value for this function. So, here's what's happening internally:

1. We create the pointer to a new MovieInfo struct.
2. We convert the body string to an array of bytes.
3. We pass the body, as bytes, and the pointer to the Unmarshal function in the json package.
   - The response of the Unmarshal function is of type error, and we don't get a real "value" returned. This is because we passed it a pointer,

so it just puts the value we expect it to "return" inside the memory that the pointer points to.

4. The function returns the pointer to the memory that we just unmarshaled within, and also, potentially, the error that the Unmarshal function may have returned.

Again, it may be a bit unintuitive to read for the first time, but once you get used to it, it starts to make a lot of sense.

Then, we simply do something similar for the function that will search for a movie by its unique identifier:

```go
func SearchById(id string) (*MovieInfo, error) {
    parms := url.Values{}
    parms.Set("apikey", APIKEY)
    parms.Set("i", id)
    siteURL := "http://www.omdbapi.com/?" + parms.Encode()
    body, err := sendGetRequest(siteURL)
    if err != nil {
        return nil, errors.New(err.Error() + "\nBody:" + body)
    }
    mi := &MovieInfo{}
    return mi, json.Unmarshal([]byte(body), mi)
}
```

The only difference here, apart from the name of the function, is on line 4 of the snippet. Instead of the parameter's name being "t" for title, it's "i" for ID.

And that's all there is to it! We now have the capability to query OMDb and get information about movies. Let's test it out with a main function:

```go
func main() {
    body, _ := SearchById("tt3896198")
    fmt.Println(body.Title)
    body, _ = SearchByName("Game of")
    fmt.Println(body.Title)
}
```

All we do in the main function is search by ID, and then search by Name. Try filling in the values here and see which movies you get! One thing to remember is that we ignore the second return values of these functions, which are both of type error. This means that if we do get an error at this point, we will not be handling it. Depending on where the error arose, this may crash the application or print out a garbage, nonsense value.

In real production code, of course, we would handle the errors more gracefully.

When you run this code, you should see the following output:

```
Guardians of the Galaxy Vol. 2
Game of Thrones
```

The first line is the title of the result with ID "tt3896198", and the second line is the title of the first result when you search for "Game of" in OMDb.

Here's the complete code listing:

CODE LISTING 3.1   Retrieve movie information using the OMDb API

```go
package main

/*
Example of only using many built-in packages in Go to reach out
to a rest API to retrieve movie detail.
*/

import (
    "encoding/json"
    "errors"
    "fmt"
    "io/ioutil"
    "net/http"
    "net/url"
    "strings"
)

//omdbapi.com API key
const APIKEY = "193ef3a"


// The structure of the returned JSON from omdbapi.com
// To keep this example short, some of the values are not
// mapped into the structure
```

```go
type MovieInfo struct {
    Title       string `json:"Title"`
    Year        string `json:"Year"`
    Rated       string `json:"Rated"`
    Released    string `json:"Released"`
    Runtime     string `json:"Runtime"`
    Genre       string `json:"Genre"`
    Writer      string `json:"Writer"`
    Actors      string `json:"Actors"`
    Plot        string `json:"Plot"`
    Language    string `json:"Language"`
    Country     string `json:"Country"`
    Awards      string `json:"Awards"`
    Poster      string `json:"Poster"`
    ImdbRating  string `json:"imdbRating"`
    ImdbID      string `json:"imdbID"`
}

func main() {
    body, _ := SearchById("tt3896198")
    fmt.Println(body.Title)
    body, _ = SearchByName("Game of")
    fmt.Println(body.Title)
}

func SearchByName(name string) (*MovieInfo, error) {
    parms := url.Values{}
    parms.Set("apikey", APIKEY)
    parms.Set("t", name)
    siteURL := "http://www.omdbapi.com/?" + parms.Encode()
    body, err := sendGetRequest(siteURL)
    if err != nil {
        return nil, errors.New(err.Error() + "\nBody:" + body)
    }
    mi := &MovieInfo{}
    return mi, json.Unmarshal([]byte(body), mi)
}
```

```go
func SearchById(id string) (*MovieInfo, error) {
    parms := url.Values{}
    parms.Set("apikey", APIKEY)
    parms.Set("i", id)
    siteURL := "http://www.omdbapi.com/?" + parms.Encode()
    body, err := sendGetRequest(siteURL)
    if err != nil {
        return nil, errors.New(err.Error() + "\nBody:" + body)
    }
    mi := &MovieInfo{}
    return mi, json.Unmarshal([]byte(body), mi)
}

func sendGetRequest(url string) (string, error) {
    resp, err := http.Get(url)
    if err != nil {
        return "", err
    }

    defer resp.Body.Close()
    body, err := ioutil.ReadAll(resp.Body)
    if err != nil {
        return "", err
    }

    if resp.StatusCode != 200 {
        return string(body), errors.New(resp.Status)
    }
    return string(body), nil
}
```

You've just seen an example of using only built-in packages to build a real, useful application. However, the true beauty of a programming language emerges when you leverage the code written by the community. To do this, we'll use Go modules.

Using Third-Party Packages

Support for Go modules began in Go version 1.11. Using modules, Go can handle third-party packages in a seamless way, enabling programmers to collaborate and share code to prevent everybody from reinventing the wheel.

Go modules can be handled by your IDE or manually via the command line. Because the purpose of this book is to learn about Go in as platform-agnostic a manner as possible, we will of course cover the way to use the command-line version.

The main command that deals with modules is "go mod". For example, if you run this in your command line:

```
go mod help
```

you should see a help page telling you all the things you can do with Go modules.

There are two paths you can take to using Go modules:

- Install a Go module "globally," meaning that you download the code for that module and store it in a path that's accessible to all of your projects. The advantage to this is that you only need to grab it once, and all of your projects suddenly have access to this module. The main disadvantage is that this can be bad sometimes, like when you need certain versions of a package or an isolated environment of packages.
- If you create your own Go module, you can install a third-party Go module within your own module. This prevents other projects and modules from accessing the module you download, and it will be stored within the project folder itself.

Let's write some code that uses a third-party Go module, and then explore both of these options. First, we recommend installing modules within your project, unless there's a very specific reason you need to install them globally.

Am I Prime?

We're going to build an application that takes a number from the user via the command line and simply prints out whether or not that number is prime. This example is quite simple, so we could easily write the code for this ourselves. However, we'll use an open source module from GitHub that does it for us.

The package we'll use can be found at www.github.com/otiai10/primes. The entire code file is quite simple:

CODE LISTING 3.2   Checking if a number is prime using otiai10's package

```go
package main

import (
    "fmt"
    "github.com/otiai10/primes"
    "os"
    "strconv"
)

func main() {
    args := os.Args[1:]
    if len(args) != 1 {
        fmt.Println("Usage:", os.Args[0], "<number>")
        os.Exit(1)
    }
    number, err := strconv.Atoi(args[0])
    if err != nil {
        panic(err)
    }
    f := primes.Factorize(int64(number))
    fmt.Println("primes:", len(f.Powers()) == 1)
}
```

As you can see, working with a third-party module is almost the same as using a built-in package! The first noticeable difference is that when you run the import, you need to specify the link to the GitHub repo (repository) where the module is hosted.

Technically, for Go modules to work, you don't need a GitHub link, just a link to any remote Git repository (perhaps hosted on GitHub, Gitlab, Bitbucket, etc.). You could even use your own Git server instance!

This enables the Go module system to handle version control. For example, it remembers specifically which commit of the module you used to compile your code, which enables more consistency in your coding environment. This information is stored within a file called "go.sum" *if* you take the second option we mentioned a moment ago, which is creating your own package.

Back to understanding our code . . . If you were to compile it at this point (using the go build . command), you'd encounter something along the lines of the following error:

```
main.go:14:2: cannot find package "github.com/otiai10/primes"
in any of:
    /usr/local/Cellar/go/1.14.4/libexec/src/github.com/otiai10/
     primes (from $GOROOT)
    /Users/tanmaybakshi/go/src/github.com/otiai10/primes
     (from $GOPATH)
```

This is Go telling you "I can't find the module you tried to import!" and so it's throwing a compiler error. If you decided on the first method of installing the package, which again we wouldn't recommend, then you'd run the following in your command line:

```
go get github.com/otiai10/primes
```

This will download the package to your home folder, so now if you build and run, the code should work normally.

However, we recommend running the following command:

```
go mod init primechecker
```

This command will do one simple thing: Within the current directory, it will create a new module called "primechecker." Within this example, the name doesn't really matter—it only comes into play when somebody else wants to use your package (this is the name they'll use to refer to your module).

Now when you run the go build command, you'll see your code compiles successfully! This is because Go is able to download the module you need into your own new module.

You should see output like the following when you run the go build command:

```
> go build .
go: finding module for package github.com/otiai10/primes go:
found github.com/otiai10/primes in github.com/otiai10/primes
v0.0.0-20180210170552-f6d2a1ba97c4
```

And you should see the following content in the go.sum file:

```
github.com/otiai10/primes v0.0.0-20180210170552-f6d2a1ba97c4/
go.mod h1:UmSP7QeU3XmAdGu5+dnrTJqjBc+IscpVZkQzk473cjM=
```

Building Your Own Packages

As we've already alluded to, in Go, making and building your own custom packages is very simple. We've even already covered how you can make your own modules, because they're important for being able to use other people's modules.

Am I Prime? Part 2

Now, in order to have others use your own custom modules, you must name your module very specifically. In our last example, you could name your module essentially anything—however, if you'd like to push this module to a remote Git server so others can use it, you must name your package as the remote Git repo that others will import in their code.

For example, let's write a package that enables a user to check if a number is prime. We initialize our package like this:

```
go mod init github.com/Tanmay-Teaches/golang/chapter3/example3
```

That way, our Go module is able to be automatically imported by people who want to use our code.

Let's write the code for this package now. Of course, we will be coding in our main.go file, but this time, instead of having "package main", we'll name the package "example3". This means that whenever people refer to this package in their own code, they'll refer to it as "example3".

CODE LISTING 3.3   Custom Prime Number Checker package

```
package example3

func IsPrime(n int) bool {
    if n <= 1 {
        return false
    } else if n <= 3 {
        return true
```

```
    } else if n % 2 == 0 || n % 3 == 0 {
        return false
    }

    i := 5
    for i * i < n {
        if n % i == 0 || n % (i + 2) == 0 {
            return false
        }
        i += 6
    }
    return true
}
```

The logic behind the IsPrime function is quite simple—we have a few conditions at the top that check for immediate cases where we can return true or false, and after that we loop from 5 to the square root of the number we want to check for. If our number is divisible by any number we encounter in our loop, then we return false; otherwise, we return true.

Now we can push the code to our GitHub repo. If this GitHub repo happens to be public, then the rest of this chapter should be quite self-explanatory, and you should be able to figure it out. All you need to do is commit your changes and push to your repo—or whatever else your Git workflow may call for.

However, you might wonder: What do we do if the GitHub repo is private? Or maybe you work at a company that uses GitHub Enterprise, so you have your own instance of GitHub. How do you tell Go to authenticate against that specific instance of GitHub?

Let's start off with figuring out how we would do that normally. Usually, you use SSH keys to authenticate against Git. SSH keys are easy to use, ultra-secure, and an industry standard.

To begin telling Go to use SSH, we actually have to start with a Git command. Specifically, we have to tell git that whenever it sees an HTTPS URL, it must replace the HTTPS request with an SSH request. This is how we can set our global Git config to do that:

```
git config --global url."git@github.com:".insteadOf
    ↪ "https://github.com/"
```

Then, you need to tell Go to *not* do a "sum check" against the private or internal repository that you're going to clone. This is because the sum check is incompatible with private and internal repositories.

```
export GONOSUMDB=github.com/Tanmay-Teaches/golang
```

That's all you need to do! At this point, you should be able to create a new project that references the module you created in a private Git repository, and then grab the module and use it in your own code. Let's go ahead and do that!

To start, create a new Go project (create a new folder) and initialize a new module, like so:

```
go mod init example4
```

Once again, when you're creating a module for your own purposes that you don't plan on distributing, the name doesn't really matter.

Apart from the module we created for ourselves, we'll use another third-party module called "echo" from "labstack". This module will let us create a quick web (HTTP) server. So, in all, our application will:

1. Open up an HTTP server at a certain port.
2. Listen for GET requests coming in at that port.
3. If it notices a GET request where the path of the URL is a number, we:
   a. Check if the number is prime.
      i. If yes, we return "true" as a response to the GET request.
      ii. If no, we return "false" as a response.

Here's all the code that goes behind this application:

CODE LISTING 3.4   Using the custom Prime Number Checker package

```
package main

import (
    "net/http"
    "github.com/Tanmay-Teaches/golang/chapter3/example3"
    "github.com/labstack/echo/v4"
    "strconv"
)
```

```go
func main() {
    e := echo.New()
    e.GET("/:number", func(c echo.Context) error {
        nstr := c.Param("number")
        n, err := strconv.Atoi(nstr)
        if err != nil {
            return c.String(http.StatusBadRequest, err.Error())
        }
        return c.String(http.StatusOK,
            ↪ strconv.FormatBool(example3.IsPrime(n)))
    })

    e.Logger.Fatal(e.Start(":1323"))
}
```

Before we run the code, let's decipher a little bit of how it works. When we start the main function, we simply call `echo.New()`, giving us a new HTTP Server instance to use. After that, we "register a GET endpoint" against that server.

We register this GET endpoint by passing it a string "/:number". This means "after the root of the URL, if the only thing you see is a certain sequence of characters that we call 'number', then yes, we can handle the request."

NOTE: When we say "number" in the string, we're not telling it to *only* look for numbers. Rather, we're saying this could be any sequence of characters, but we want to store those characters in a variable called "number", because the API we're designing expects this to be a number.

However, there is one more argument that we pass to this function: It's another function, which is not immediately invoked. That means after the "func" keyword on line 2 of the main function, until line 9 of the main function, that code is *not* immediately executed—it's just part of a new function that we're declaring inline and passing to the GET function from the echo module.

This function we pass will be invoked by the HTTP server whenever it finds a request that fulfills the condition made by the string we passed in as the first argument to the GET function. When it's invoked, it'll be passed an `echo.Context` value, and is expected to return an error type. This will be `nil` if there is no error. If there *is* an error, which means the return value is not `nil`, the echo module will handle sending a failure response to the request for you.

Within this function, we simply get the "number" parameter we declared in the path string we passed. We then try to convert it to an integer. If there is an error at this point, meaning that the string really contained something other than just an integer, we return that error, along with some extra information that tells echo "the user of the API made a bad request, so we can't handle it. It's not our fault!"

Finally, we simply tell the server to start listening for requests on port 1323. This function will return an error if there is one, and if there is no error, it will *never* return, it will block. This is normal server code behavior. When you force kill the program, it will end.

However, if the function *does* return an error, we pass that error into the Fatal function within Logger. This is so we can log the error as a "fatal error," and because the main function then ends, the program will exit.

That's enough talking. Let's see the code in action! If you go to your browser, you can hit the endpoint with numbers, as you can see in Figures 3.1 and 3.2.

false

FIGURE 3.1   The output of the Prime Number Checker API for 100, a composite number

true

FIGURE 3.2   The output of the Prime Number Checker API for 97, a prime number

For each request, you should see "true" or "false" as output: true if the number was prime, and false if the number was composite.

In this chapter, we explored how to use Go modules—both third-party and built-in—and we showed you how you can build your own! In the chapters to come, we will leverage all the concepts you've learned so far, in order to build real applications that use common data structures and algorithms.

Exercises

1. What is a "checksum" and why does Go use it? In which cases should you disable it?

2. If you import a module in Go, why does the compiler throw an error if it's not used?

3. Why is it preferred to install Go modules separately for each project instead of globally?

4. What naming scheme should you use for packages that you plan to redistribute as custom modules on a version control system?

5. Why can you not return nil from a function if the return type is not a pointer?

6. How do you stop an HTTP server after the blocking `start` function is called?

Using Built-in Packages

Welcome to Chapter 4! Now that you've been introduced to the basic building blocks of the Go programming language and how you can use packages and modules, it's time to learn some additional features and capabilities of the language. We'll do this by building applications powered by some common data structures and algorithms you'll likely use in your own apps.

Once you've finished this chapter, you'll be able to answer the following questions:

- What is the Dijkstra pathfinding algorithm and how does it work?
- How can you implement Dijkstra to search through a graph in Go?
- What is Conway's Game of Life and how does it work?
- How can you implement graphical user interfaces in Go?
- Why is generating random numbers a difficult task and how can you build your own pseudorandom number generator in Go?

Common Data Structures and Algorithms

Specifically, we'll take a look at a couple of famous algorithms that can teach us a lot about transforming pseudocode and logic into Go code. These two examples are:

- Dijkstra's pathfinding
- Conway's Game of Life

As we do this, we'll not only revise Go concepts and learn new ones, we'll also move onto a whole new paradigm of user interaction we haven't covered yet: graphical user interfaces.

Go as a language was never built to work on the UI side of programming. However, it is definitely possible to use Go to build GUIs. We'll be diving further into this in a few moments.

Dijkstra's Pathfinding

Let's start with the algorithm you've likely heard a lot more about: Dijkstra's pathfinding algorithm. As you may know, a pathfinding algorithm lets you find a "route" from one node to another in a graph.

The word "graph" might seem a bit limiting, but lots of things in the world can be turned into a "graph." Take, for example, a maze. You can represent the maze as a graph of individual tiles that connect to each other, where some connections are blocked by walls.

Or you could even take your city's road system. Different roads are nodes that are connected to each other with a certain "cost," which in the real world usually translates to "distance." However, by using particular technologies, you can factor other features into this cost as well, such as traffic, the straightforwardness of a route, and others.

For this example, let's say we're working on the graph shown in Figure 4.1.

What we want to do is figure out that, when given any two nodes on this graph, what route should we take to move from the first node to the second node that incurs the least cost? For example, to move from C to D, you could do C → B → D, but the cost would add up to 7, because C → B is 5 and B → D is 2. Therefore, a smarter route is C → E → D, in which case the cost is 6, because C → E is 5 and E → D is 1.

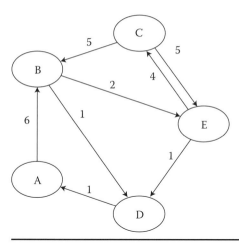

FIGURE 4.1   Sample graph to run Dijkstra pathfinding on

The way Dijkstra works is interesting in the sense that it looks at every possible path instead of trying to narrow down its search to paths that are more likely to move toward our goal. In certain extensions of this algorithm, such as the A* algorithm, we use what's known as a "heuristic" in order to guide the technique to make better choices, and to therefore be more efficient. A heuristic is an imperfect "estimate" of how close a node is to our eventual goal.

The inner working of Dijkstra is deceptively simple—so much so that it actually might take you a moment to comprehend that the algorithm's behavior might end up allowing us to solve pathfinding! The best way to really understand the algorithm is to visualize it step by step. So, let's use a simplified version of the graph in Figure 4.1, which you can see in Figure 4.2, and work through the Dijkstra algorithm manually first:

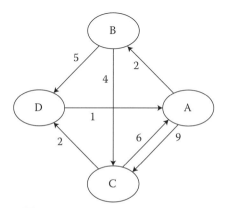

FIGURE 4.2   A simpler graph to practice Dijkstra's algorithm

When the algorithm starts, we create two dictionaries where the keys are the nodes within the graphs, and the values are floats and nodes, respectively. These two dictionaries are called the "distance" or "cost" and the "parent" or "previous" dictionaries. We'll get into what those mean in just a moment. First, let's take a look at a tabular representation of those dictionaries in Table 4.1.

TABLE 4.1   Initial Dijkstra Pathfinding State

NODE	COST	PARENT
A	INF	nil
B	INF	nil
C	INF	nil
D	INF	nil

As you can see, the default value for the cost dictionary is infinity, and the default value for the parents is nil.

When the algorithm starts, we need to know the source node that we'll start the pathfinding from. For this node, we will set the cost to 0—intuitively, this makes sense. The cost to get from the source to the source is 0, because you're already there! The parent is also nil—because how did you get to the source? You didn't get there *from* the source, you were just already there; no node precedes it.

Let's say, for the sake of discussion, the source node in our graph is node B. The dictionaries should look like Table 4.2 now.

TABLE 4.2   Dijkstra Pathfinding State After Updating the Root Node's Cost

NODE	COST	PARENT
A	INF	nil
B	0	nil
C	INF	nil
D	INF	nil

All we need to do now is run a normal Dijkstra's algorithm iteration. We start by looking at the costs of all the nodes and finding the node with the smallest cost. Because all of them apart from B are infinite, of course, B is the node with the lowest cost.

Then, we query the graph to find out what the neighbors of that node are, and what their costs are. We see that the neighbors in this specific graph are nodes C and D.

And yes, technically, B shares a connection with A, but that's a connection FROM A to B, so we can't move from B to A, and that connection is therefore invalid for this logic.

We start with node C (the order does not matter in this case). We check, is the cost of B PLUS the cost it would take to travel to C lower than the cost of C itself? In this case, that means: "Is 0 + 4 less than infinity?" Of course, this resolves to true. Therefore, we set the cost of C to 4 (because 0 + 4 = 4), and we set the parent of C to B.

We do something similar with node D. Because the cost was previously infinity, the new cost of 0 (B's cost) plus 5 (the cost of moving from B to D) is lower—because any finite number is lower than infinity. So we set the cost to 5 and set the parent to B.

There we go! We've iterated through the movements you could make from B. Let's run the iteration again. Table 4.3 is what our dictionaries look like at the moment.

TABLE 4.3   Dijkstra Pathfinding State After Updating C and D Nodes' Costs and Parents

NODE	COST	PARENT
A	INF	Nil
B	0	Nil
C	4	B
D	5	B

Looking at the table, we find that C currently has the lowest cost. So, let's see what outbound connections C has. We see that C can move to A and D. Let's start with determining its relationship with A.

Node A currently has a cost of infinity, so regardless of the cost of C + (C → A), it will be lower. In this case, that resolves to 10, because C is 4, and the connection cost is 6. Therefore, we set the cost of A to 10 and the parent to C.

Then, we look at D. D already has a cost of 5. C has a cost of 4, as we've seen, and the connection between C and D has a cost of 2. So, 4 + 2 = 6, which means that the connection between C and D would be more expensive than the connection that D already has, which is from B. Therefore, we don't modify D's values in the dictionaries at all.

Table 4.4 is the current state of our dictionaries.

TABLE 4.4   Dijkstra Pathfinding State After Updating Node A's Cost and Parent

NODE	COST	PARENT
A	10	C
B	0	Nil
C	4	B
D	5	B

But we're not done yet! Remember that, so far, we've run the algorithm's iteration on nodes B and C. So, if we look for the node with the lowest cost in the table that we have *not* yet iterated on, we see that node is D. Let's go ahead and run an iteration on D.

D can connect only to A. Let's try and determine a new cost for A if we were to switch its parent to D. D's cost is 5, and the cost of its connection to A is only 1. $5 + 1 = 6$, which is *lower* than the cost for A that's already in the table, which is 10. Therefore, we change A's cost from 10 to 6 and A's parent from C to D.

After this update, our dictionaries look like Table 4.5.

TABLE 4.5   Dijkstra Pathfinding State After Updating Node A's Cost and Parent

NODE	COST	PARENT
A	6	D
B	0	Nil
C	4	B
D	5	B

All right, the only node left is A. Let's try running an iteration on it. A connects to both B and C.

Should we replace B's cost and parent? Intuitively, we know the answer is no because B is the source node, it makes no sense to make A its parent. However, we can also prove this with the algorithm, because no matter what the cost is for the connection between A and B, we will never replace B's parent, because B's cost is already as low as we can get it: zero. So, let's leave B alone.

What about C? A's cost is already 6, plus the connection from A → C, which is 9. That adds up to 15. 15 is greater than the existing cost of 4, therefore we leave C alone as well. This means the pathfinding state doesn't change; it remains the same as in Table 4.5.

Now, finally, how in the world does this tell us how to move from B to any other node? The answer is hidden in plain sight in this table. What Dijkstra did is it created ANOTHER graph where we know what the cheapest preceding node is to come in from for every node, considering you originate at a certain source.

For example, let's say you want to go from B to A. Let's look at the table. What's the parent for node A? It's node D. What's the parent for node D? It's node B. But hey! Node B is our origin—so we can stop our loop.

This means B → D → A is the most efficient way to get from B → A.

Once this algorithm clicks in your head, the code makes a *lot* more sense. Otherwise, the abstraction layer created by the code makes it more difficult to really understand why this table of the most efficient path to any node emerges.

Now let's code this in Go! Before we touch any real code, let's take a look at some pseudocode for Dijkstra's algorithm:

```
function Dijkstra(Graph, source):

    create vertex set Q

    for each vertex v in Graph:
        dist[v] ← INFINITY
        prev[v] ← UNDEFINED
        add v to Q
    dist[source] ← 0

    while Q is not empty:
        u ← vertex in Q with min dist[u]

        remove u from Q

        for each neighbor v of u: // only v that are still in Q
            alt ← dist[u] + length(u, v)
            if alt < dist[v]:
                dist[v] ← alt
                prev[v] ← u

    return dist[], prev[]
```

As you can see, the pseudocode is quite similar to the process we described earlier.

When it comes to implementing this algorithm in Go, there are various different ways of doing so. We'll stick to the method that stays true to the logic we introduced you to a moment ago.

We know there is some basic information we need to have for this to work, such as how nodes are connected to each other, the cost of moving from one node to another, and the directions of the connections. In order to encode this information in Go, we'll use the following three structures:

```go
type Node struct {
    Name  string
    links []Edge
}


type Edge struct {
    from *Node
    to   *Node
    cost uint
}


type Graph struct {
    nodes map[string]*Node
}
```

The "Node" structure contains the information for a single node, which is basically the other nodes it connects to, plus the name of the node.

The "Edge" structure contains a pointer to the node it comes from, a pointer to the node it goes to, and the cost of deciding to take that connection.

The "Graph" structure only contains a single value, which is the map of node names to the pointers to the nodes themselves.

We also define a very small helper function to let us create a new Graph structure instance and get a pointer to it:

```go
func NewGraph() *Graph {
    return &Graph{nodes: map[string]*Node{}}
}
```

Before we can start coding the fun pathfinding, we need to write some infrastructure, so users are able to interact with the actual pathfinding algorithm.

This means they must be able to take a graph of theirs and somehow represent it using the three structures we defined earlier.

We'll use the following two functions to help them do that:

```
func (g *Graph) AddNodes(names ...string) {
    for _, name := range names {
        if _, ok := g.nodes[name]; !ok {
            g.nodes[name] = &Node{Name: name, links: []Edge{}}
        }
    }
}


func (g *Graph) AddLink(a, b string, cost int) {
    aNode := g.nodes[a]
    bNode := g.nodes[b]
    aNode.links = append(aNode.links, Edge{from: aNode,
        ↪ to: bNode, cost: uint(cost)})
}
```

These functions are mostly self-explanatory. We add two pointer receiver functions to the Graph type, one for adding new nodes with no links, and one for adding links to existing nodes.

If you want to add nodes, all you do is, for each node you create a new node, assign it a name and no edges, get a pointer to that node, and assign that pointer as a value in the nodes map within the graph you called the function on.

If you'd like to add a link, all you do is get the nodes from the graph, and append a new edge to the links of the first node.

Now let's implement the actual pathfinding! First, how do we represent the costs and nodes? Well, we already have a structure for the nodes, and for the sake of simplicity, we'll use the uint datatype for the costs instead of float. This raises an issue though: How can we represent infinity with an integer?

Floating point values have a certain set of bits that represent the infinity value, but the same is not true for integers. Therefore, to represent infinity, we'll simply use the maximum value an unsigned integer can hold in Go. So, create a global constant containing this max value:

```
const INFINITY = ^uint(0)
```

This works by taking an unsigned integer with the value of 0 — meaning all the bits in the number are 0—and running a bitwise NOT operator on them, which is the carat symbol. This turns every bit to 1, bringing the value of the variable to the maximum an unsigned integer can hold.

Time to implement Dijkstra's algorithm. In the following we have the same input and output as the pseudocode we saw a little while ago. It only needs to know the source node, and then it calculates the costs and parents for you:

```go
func (g *Graph) Dijkstra(source string) (map[string]uint,
    ↪ map[string]string) {
    dist, prev := map[string]uint{}, map[string]string{}

    for _, node := range g.nodes {
        dist[node.Name] = INFINITY
        prev[node.Name] = ""
    }
    visited := map[string]bool{}
    dist[source] = 0
    for u := source; u != ""; u = getClosestNonVisitedNode(dist,
        ↪ visited) {
        uDist := dist[u]
        for _, link := range g.nodes[u].links {
            if _, ok := visited[link.to.Name]; ok {
                continue
            }
            alt := uDist + link.cost
            v := link.to.Name
            if alt < dist[v] {
                dist[v] = alt
                prev[v] = u
            }
        }
        visited[u] = true
    }
    return dist, prev
}
```

```go
func getClosestNonVisitedNode(dist map[string]uint,
    ↪ visited map[string]bool) string {
    lowestCost := INFINITY
    lowestNode := ""
    for key, dis := range dist {
        if _, ok := visited[key]; dis == INFINITY || ok {
            continue
        }
        if dis < lowestCost {
            lowestCost = dis
            lowestNode = key
        }
    }
    return lowestNode
}
```

To take a look at how this code works, we think it's best to start at the very end of the code snippet, with the second function in the code. As you know, the pseudocode only had a single function defined, so why do we have two separate functions in this snippet for pathfinding?

It's because Go is, as we mentioned, a language that doesn't give you many luxuries. There are lots of little bits of functionality that you'd expect to have, that you unfortunately need to implement yourself.

In some cases, this can be really good for performance because instead of using the standard library for everything, and designing data structures around being able to use the library, you instead design algorithms around your data. However, this can also sometimes be detrimental for performance, because you need to reinvent the wheel quite often and implement functionality that could've been done better by the open source community.

In this case, the second function has a very simple task: Take the dictionary that tells us the "costs" of the nodes (in this case, "distances"), as well as a map that tells us whether or not a node has been visited, and figure out which node has the lowest cost and also has *not* been visited yet.

The function does this by creating two variables: the distance of the node that has the lowest distance value, and the name of that node. At first, we don't know these values, so we simply fill placeholder values. For example, the distance is infinite—you'll see why in a moment.

Then, we loop through the different values in the `dist` (distances) map, and when we see a node with a distance lower than the one we've stored in the `lowestNode` and `lowestCost` variables, we update the lowest distance and also the name of the node that has that distance.

Getting back to why the initial value for the distance is infinite . . . It's so that no matter what the value of the first non-infinite distance we encounter is, we end up replacing the placeholder value, because any number that's not infinity itself will be lower than infinity.

The logic on line 4 of the second function (`getClosestNonVisitedNode`) in the snippet is quite interesting, so it's worth expanding upon before we continue:

```
if _, ok := visited[key]; dis == INFINITY || ok {
```

This line of code tries to query the "visited" dictionary with the key that we got from the distances dictionary. When the dictionary returns its two response values, we ignore the first one, which is the actual value, and we instead only focus on the second response from the dictionary. This second response tells us whether or not that key even exists in the dictionary. If the key does exist, this node was visited (regardless of its value in the dictionary), and we skip this iteration of the loop. We also skip this iteration if we know that the cost of this node is infinite, meaning it hasn't been calculated or set yet.

Now, we can finally focus on the first function, which is where the real fun is. It's implemented as a pointer receiver on the Graph structure. This is because we may need to modify the graph itself within this function.

Let's recap, section by section, what the function does:

```
dist, prev := map[string]uint{}, map[string]string{}

for _, node := range g.nodes {
    dist[node.Name] = INFINITY
    prev[node.Name] = ""
}
visited := map[string]bool{}
dist[source] = 0
```

This first section starts by creating the cost and parent (`dist` and `prev`) dictionaries and initializes them in the way that was described at the beginning of the chapter. The costs are all infinite and the parents are all nil. (In this case, nil is represented as an empty string, because the value is not a pointer.) We also create the dictionary that represents whether or not a node has been visited, and we set the distance, which is the cost, for the source node to 0.

```
for u := source; u != ""; u = getClosestNonVisitedNode(dist,
    ↪ visited) {
  uDist := dist[u]
```

This section of the code has one simple responsibility: start a for loop with the variable u that starts with the source string, and then loops until the u variable is equal to an empty string. After every iteration, call the getClosestNonVisitedNode function, pass it the dictionaries it needs, and store the result in u. The logic behind this line is that when this function returns an empty string, we've already visited all the nodes and can stop the loop.

After we start the loop itself, we simply query the cost and distance dictionary for the node that we're on and store that distance in uDist.

```
for _, link := range g.nodes[u].links {
    if _, ok := visited[link.to.Name]; ok {
        continue
    }
    alt := uDist + link.cost
    v := link.to.Name
    if alt < dist[v] {
        dist[v] = alt
        prev[v] = u
    }
}
visited[u] = true
```

This entire section is essentially just responsible for going through the link this node has outbound, and running the Dijkstra iteration we talked about earlier. It starts by checking if this node has already been visited, in which case we refrain from running the iteration.

If the node has not been visited, then we do run the iteration, which consists of calculating a new proposed cost for this node, checking if the proposed cost is lower than the previously calculated cost, and if so, updating the node's parent and cost values.

After running the iterations, we also make sure to jot down that we just visited the node stored in u, so that we don't run the iteration on this node again.

After that section, only one task remains! We need to return the cost and parent dictionaries, so the caller of the function can run the actual pathfinding if they need it:

```
return dist, prev
```

Just before we start using this code, we'll define a short helper function that prints out the values from these dictionaries in a pretty way:

```
func DijkstraString(dist map[string]uint,
    ↪ prev map[string string) string {
    buf := &bytes.Buffer{}
    writer := tabwriter.NewWriter(buf, 1, 5, 2, ' ',0)
    writer.Write([]byte("Node\tDistance\tPrevious Node\t\n"))
    for key, value := range dist {
        writer.Write([]byte(key + "\t"))
        writer.Write([]byte(strconv.FormatUint(uint64(value),
            ↪ 10) + "\t"))
        writer.Write([]byte(prev[key] + "\t\n"))
    }
    writer.Flush()
    return buf.String()
}
```

You don't really need to worry about how this function works, but on a high level, the inputs are the two dictionaries from the Dijkstra function, and the output is a string that represents a pretty, tabular version of the dictionaries.

This function works by creating a new `Buffer` of bytes, which will eventually be converted to our string. We use a module called `tabwriter` to make this a lot easier for us. We call `NewWriter` on `tabwriter` to get a new instance of a `writer`, with a minimum width of 1, a tab width of 5, and padding of 2 characters: the padding character being a space, and no special flags. We then write the header row, followed by each key in the `dist` (cost) dictionary, as well as its value, and then its parent from the `prev` dictionary.

We then call `Flush()` on `writer` (telling it to move whatever changes are cached to the actual buffer), then we convert the buffer to a string and return that string from the function.

Now, let's see how we can use these dictionaries to run pathfinding on a real graph. Let's say we're working on the graph from earlier, in Figure 4.3.

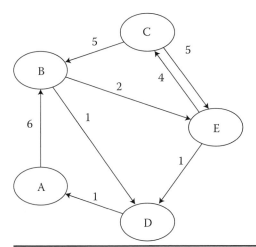

FIGURE 4.3 The sample graph to use for Dijkstra pathfinding

Now let's write a main function for this program. The main function will simply re-create the preceding graph within the structures we defined at the beginning of our code, and then run Dijkstra from a certain source node.

```
func main() {
        g := NewGraph()
        g.AddNodes("a", "b", "c", "d", "e")
        g.AddLink("a", "b", 6)
        g.AddLink("d", "a", 1)
        g.AddLink("b", "e", 2)
        g.AddLink("b", "d", 1)
        g.AddLink("c", "e", 5)
        g.AddLink("c", "b", 5)
        g.AddLink("e", "d", 1)
        g.AddLink("e", "c", 4)
        dist, prev := g.Dijkstra("a")
        fmt.Println(DijkstraString(dist, prev))
}
```

Your code should look like the following now:

CODE LISTING 4.1   "Dijkstra's pathfinding" application

```go
package main

import (
    "bytes"
    "fmt"
    "strconv"
    "text/tabwriter"
)

const INFINITY = ^uint(0)

type Node struct {
    Name  string
    links []Edge
}

type Edge struct {
    from *Node
    to   *Node
    cost uint
}

type Graph struct {
    nodes map[string]*Node
}

func NewGraph() *Graph {
    return &Graph{nodes: map[string]*Node{}}
}

func (g *Graph) AddNodes(names ...string) {
    for _, name := range names {
        if _, ok := g.nodes[name]; !ok {
```

```go
        g.nodes[name] = &Node{Name: name, links: []Edge{}}
    }
  }
}


func (g *Graph) AddLink(a, b string, cost int) {
    aNode := g.nodes[a]
    bNode := g.nodes[b]
    aNode.links = append(aNode.links, Edge{from: aNode,
        ↪ to: bNode, cost: uint(cost)})
}
func (g *Graph) Dijkstra(source string) (map[string]uint,
    ↪ map[string]string) {
    dist, prev := map[string]uint{}, map[string]string{}

    for _, node := range g.nodes {
        dist[node.Name] = INFINITY
        prev[node.Name] = ""
    }
    visited := map[string]bool{}
    dist[source] = 0
    for u := source; u != ""; u = getClosestNonVisitedNode(dist,
        ↪ visited) {
        uDist := dist[u]
        for _, link := range g.nodes[u].links {
            if _, ok := visited[link.to.Name]; ok {
                continue
            }
            alt := uDist + link.cost
            v := link.to.Name
            if alt < dist[v] {
                dist[v] = alt
                prev[v] = u
            }
        }
        visited[u] = true
    }
```

```go
        return dist, prev
}


func getClosestNonVisitedNode(dist map[string]uint,
    ↪ visited map[string]bool) string {
    lowestCost := INFINITY
    lowestNode := ""
    for key, dis := range dist {
        if _, ok := visited[key]; dis == INFINITY || ok {
            continue
        }
        if dis < lowestCost {
            lowestCost = dis
            lowestNode = key
        }
    }
    return lowestNode
}


func main() {
        g := NewGraph()
        g.AddNodes("a", "b", "c", "d", "e")
        g.AddLink("a", "b", 6)
        g.AddLink("d", "a", 1)
        g.AddLink("b", "e", 2)
        g.AddLink("b", "d", 1)
        g.AddLink("c", "e", 5)
        g.AddLink("c", "b", 5)
        g.AddLink("e", "d", 1)
        g.AddLink("e", "c", 4)
        dist, prev := g.Dijkstra("a")
        fmt.Println(DijkstraString(dist, prev))
}


func DijkstraString(dist map[string]uint,
    ↪ prev map[string]string) string {
    buf := &bytes.Buffer{}
```

```
writer := tabwriter.NewWriter(buf, 1, 5, 2, ' ',0)
writer.Write([]byte("Node\tDistance\tPrevious Node\t\n"))
for key, value := range dist {
    writer.Write([]byte(key + "\t"))
    writer.Write([]byte(strconv.FormatUint(uint64(value),
        ↪ 10) + "\t"))
    writer.Write([]byte(prev[key] + "\t\n"))
}
writer.Flush()
return buf.String()
```
}

In this case, the starting node is "a". So, you see the following output when you run the code to pathfind outward from "a" to every other node:

```
Node   Distance   Previous Node
a      0
b      6          a
c      12         e
d      7          b
e      8          b
```

So now, if you want to determine the best way to get from A to D, all you need to do is follow the dictionary: D's parent is B, B's parent is ... A! So A → B → D is the cheapest way to get from A to D.

Although this is a very simple example to demonstrate how you can use Dijkstra's algorithm to ascertain the shortest path in a graph, this same code can be used in larger scale applications. Plus, the concepts learned during the implementation of this demo will help you in your journey of applying Go.

Conway's Game of Life

Conway's Game of Life is quite a popular game among people in the math and technology communities. It's not like your traditional game—it's special, in that it's a *zero-player* game. That means you don't get to modify the environment or state after you set it for the first time.

It was invented by mathematician John Conway back in 1970. Technically, the Game of Life is a kind of "cellular automaton," meaning the game consists of a collection of cells which, according to some mathematical rules, will either live, die, or multiply.

What's really interesting is you can get some very complex behavior to emerge from a very simple initial state. Take the following grid in Figure 4.4, for example.

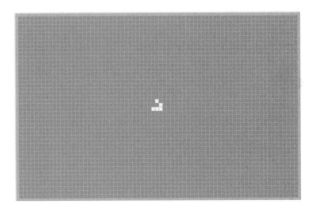

FIGURE 4.4   Sample "Conway's Game of Life" grid at the beginning of the game

This is a grid, which is really a collection of cells. Every cell that's gray is "dead," and all the yellow cells are "alive." At each iteration, you use a few rules to determine if a cell should die, become alive, or stay the same. With this specific initial setup, using Conway's rules, you should see the yellow cells drift to the bottom right of the grid infinitely.

Looking at static images of the Game of Life in action really cannot do it justice. It's absolutely worth it to go to the online implementation of this game at Edwin Martin's website, www.bitstorm.org/gameoflife.

Now, let's take a look at how we can implement this game in Go! But first, let's quickly go over our rules:

1. If a cell is already alive
   a. If it has less than 2 living neighbors, it dies.
   b. If it has 2 or 3 living neighbors, it stays alive.
   c. If it has more than 3 living neighbors, it dies.
2. If a cell is already dead
   a. If it has exactly 3 living neighbors, it becomes alive.

The rules are deceptively simple—and not only do they emerge in visually interesting behavior, they're even *Turing complete*! You could implement a calculator, the Go programming language, or any computable task in a single, initial Game of Life state.

In this example application, we'll do things a bit differently. We'll build a graphical user interface, so you can actually see what the Game of Life is doing. For this, we'll need a couple of extra imports. Specifically, we'll use a package called "pixel." The following are all the imports we'll need:

```
import (
    "github.com/faiface/pixel"
    "github.com/faiface/pixel/pixelgl"
    "image/color"
    "math/rand"
)
```

We'll also create a structure that's responsible for storing the information of what we're rendering to the screen. It stores only two pieces of very simple information: the pixels that we're rendering, and the width of the screen. The reason we only store width and not height is because we don't need it to calculate the index of the pixels. We'll explore this in a moment.

One thing to note: The way the pixels are stored is in a single, one dimensional array of unsigned 8-bit integers. Each pixel has four values in this array, in the following order: Red, Green, Blue, Alpha. You're likely familiar with RGB, and Alpha means transparency.

```
type Pixels struct {
    //RGBA colors
    Pix    []uint8
    Width int
}
```

Before we continue, understand that we don't store the height. This is because when we need to find the index of the pixel in the Pix array given a certain (x, y) coordinate (here, x and y are indices from the top-left corner of the screen, for which the coordinates are (0,0)), all you need to do is figure out $((y * width) + x) * 4$, and you get the index of the R element of the pixel. Plus 1, you get G; plus 2, you get B; plus 3, you get A; and plus 4, you've found the R element of the next pixel.

Therefore, while we could store the height for safety reasons, such as to make sure the user gave us a valid Y coordinate, we don't *need* to.

Now we need to create some functions to help us interface with this structure more easily—for example, initializing the structure to a screen of just black pixels with a certain width, or a function that can set the color of a pixel at a certain coordinate.

Technically, we could use the functions already built for this task within the Pixel module that we've imported. However, graphics rendering is compute intensive, and the Pixel module has a relatively slow implementation of these methods. If we were to use those methods directly, it would be impossible to run a large-scale Game of Life in real time. It would take a couple of seconds, not to compute the iteration, but for Pixel to render the iteration that we compute!

Therefore, we'll write our own function to implement these basic pieces of functionality. In a way, this is an opportunity for you to learn a few new tricks and to get in the habit of learning how to solve problems that come your way, even if they're caused by other people's code! Remember, Go was never really built for GUI work, but if you *really* need to, you *can* use it.

```go
func NewPixels(width, height int) *Pixels {
    return &Pixels{Width: width, Pix: make([]uint8,
        ↳ width*height*4)}
}


func (p *Pixels) DrawRect(x, y, width, height int,
    ↳ rgba color.RGBA) {
    for idx := 0; idx < width; idx++ {
        for idy := 0; idy < height; idy++ {
            p.SetColor(x+idx, y+idy, rgba)
        }
    }
}


func (p *Pixels) SetColor(x, y int, rgba color.RGBA) {
    r, g, b, a := rgba.RGBA()
    index := (y*p.Width + x) * 4
```

```
    p.Pix[index] = uint8(r)
    p.Pix[index+1] = uint8(g)
    p.Pix[index+2] = uint8(b)
    p.Pix[index+3] = uint8(a)
}
```

As you can tell, these functions are quite simple. `NewPixels` will simply initialize the `pixels` array, create a new `Pixels struct`, and pass us a pointer to it. `DrawRect` will take the top-left coordinate of a rectangle, the `width` and `height`, and the color, and draw that rectangle into the `Pixels structure`. Finally, `SetColor` will find the 1D (one-dimensional) index of a 2D coordinate to a pixel and set the color accordingly.

All right! We've gotten most of the GUI infrastructure out of the way. Now we can work on the actual Game of Life logic. As we've already explored, one of the main tasks in the Game of Life is being able to count the neighbors of a certain node. Unlike some other algorithms, even nodes at a diagonal are considered to be valid neighbors in the Game of Life. Therefore, in Figure 4.5, for the node in the middle (the node with a circle), all the eight nodes pointed to with arrows are considered neighbors.

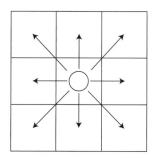

FIGURE 4.5   The boxes that count as a "neighbor" for one box

In order to implement the logic that counts the number of neighbors that are still alive, we'll start from the top left of the neighbors, move to the end of that row, then move down a row, move to the end of that row, and so on. We will skip any indices that would be off the screen (such as for a box at the very left of the screen; we won't count neighbors further to the left). We will also skip the node that we're counting neighbors for.

Here's the function implementing that logic:

```go
func CountNeighbors(matrix [][]int) [][]int {
    neighbors := make([][]int, len(matrix))
    for idx, val := range matrix {
        neighbors[idx] = make([]int, len(val))
    }
    for row := 0; row < len(matrix); row++ {
        for col := 0; col < len(matrix[row]); col++ {
            for rowMod := -1; rowMod < 2; rowMod++ {
                newRow := row + rowMod
                if newRow < 0 || newRow >= len(matrix) {
                    continue
                }
                for colMod := -1; colMod < 2; colMod++ {
                    if rowMod == 0 && colMod == 0 {
                        continue
                    }
                    newCol := col + colMod
                    if newCol < 0 || newCol >= len(matrix[row]) {
                        continue
                    }
                    neighbors[row][col] += matrix[newRow][newCol]
                }
            }
        }
    }
    return neighbors
}
```

If you look closely, you'll see precisely the logic that moves from the top left down each row one by one to count neighbors. The first for loop in the function is simply responsible for initializing a two-dimensional array that will store the number of "alive" neighbors for each cell.

Then, the second for loop, which is really just the first for loop in an enclosure of three total, loops through every single cell, running the logic to compute the number of "alive" neighbors for that cell using the inner two loops.

The first of the four nested loops (the outermost loop) is responsible for moving from the beginning to the end of a row. The second inner loop is responsible for moving from one row to another. Whenever we encounter a row or column that would be impossible to get the value for, like something that's out of bounds, we skip the iteration. If we find the row and column have the same index as the cell we're trying to calculate neighbors for, we also skip it.

However, if none of these guard conditions is hit, and we are at a valid index, then we look at the value of the neighbor, and if it's alive, we increment the value of this cell in the "neighbors" array.

Before we continue, let's create another structure called GameOfLife, which contains a game board that represents the current state of the environment, the pixels we'll render of that game board, and also the size of the cells on the board itself. This is because a single cell won't just be a single pixel, that would be incredibly tiny—instead, each cell will have a certain "size" on the screen, and that's what we need to store.

```
type GameOfLife struct {
    gameBoard [][]int
    pixels    *Pixels
    size      int
}
```

To add to this, let's create some simple helper functions that can create a new Game of Life instance with an empty board, as well as a function that can populate the board with random cells, meaning they all have a 50/50 chance of being dead or alive:

```
func NewGameOfLife(width, height, size int) *GameOfLife {
    gameBoard := make([][]int, height)
    for idx := range gameBoard {
        gameBoard[idx] = make([]int, width)
    }
    pixels := NewPixels(width*size, height*size)
    return &GameOfLife{gameBoard: gameBoard, pixels: pixels,
        ↪ size: size}
}

func (gol *GameOfLife) Random() {
```

```
        for idy := range gol.gameBoard {
            for idx := range gol.gameBoard[idy] {
                gol.gameBoard[idy][idx] = rand.Intn(2)
            }
        }
    }
```

Now let's create a function that can help us actually run a single iteration of Conway's Game of Life according to the conditions we defined earlier on in this section and the output from the countNeighbors function that we just defined.

Within this function, we also update the canvas itself. Because we're just updating the canvas that we've created ourselves, and we're not actually rendering out to the screen, it's super-fast and adds barely any overhead.

```
func (gol *GameOfLife) PlayRound() {
    neighbors := CountNeighbors(gol.gameBoard)
    for idy := range gol.gameBoard {
        for idx, value := range gol.gameBoard[idy] {
            n := neighbors[idy][idx]
            if value == 1 && (n == 2 || n == 3) {
                continue
            } else if n == 3 {
                gol.gameBoard[idy][idx] = 1
                gol.pixels.DrawRect(idx*gol.size, idy*gol.size,
                    ↪ gol.size, gol.size, Black)
            } else {
                gol.gameBoard[idy][idx] = 0
                gol.pixels.DrawRect(idx*gol.size, idy*gol.size,
                    ↪ gol.size, gol.size, White)
            }
        }
    }
}
```

This code is also quite simple. It implements a pointer receiver to the GameOfLife structure, takes no extra arguments, and returns nothing because it modifies the original structure it was passed a pointer to.

Within the function, we count all the "alive" neighbors for each cell in the game board, and we then loop through every Y index in the game board. We subscript the board at every iteration to get that row, and loop through every column in that row. If the cell is alive and the neighbor count is 2 or 3, we keep the cell as it is. Or else, if the neighbor count is 3, we turn the cell alive. Once again, if none of the previous conditions evaluated to true, we turn the cell dead.

Whenever we change the cell's living status to alive or dead, also make the necessary change in the pixels.

Finally, let's tie all of this together into a main function that creates a canvas on which to render pixels, creates the Game of Life, runs the iterations, and updates the canvas. The iterations of the Game of Life will end and the program will exit once the window is closed.

```go
func run() {
    size := float64(2)
    width := float64(400)
    height := float64(400)
    cfg := pixelgl.WindowConfig{
        Title:  "Conway's Game of Life",
        Bounds: pixel.R(0, 0, width*size, height*size),
        VSync:  true,
    }
    win, err := pixelgl.NewWindow(cfg)
    if err != nil {
            panic(err)
    }
    gol := NewGameOfLife(int(width), int(height), int(size))
    gol.Random()
    for !win.Closed() {
        gol.PlayRound()
        win.Canvas().SetPixels(gol.pixels.Pix)
        win.Update()
    }
}


func main() {
    pixelgl.Run(run)
}
```

Your code should now look like this:

CODE LISTING 4.2   "Conway's Game of Life" application

```go
package main

import (
    "github.com/faiface/pixel"
    "github.com/faiface/pixel/pixelgl"
    "image/color"
    "math/rand"
)

var (
    Black = color.RGBA{0, 0, 0, 255}
    White = color.RGBA{255, 255, 255, 255}
)

//Canvas
type Pixels struct {
    //RGBA colors
    Pix    []uint8
    Width int
}

//Create a new canvas with dimension width x height
func NewPixels(width, height int) *Pixels {
    return &Pixels{Width: width, Pix: make([]uint8,
        ↪ width*height*4)}
}

func (p *Pixels) DrawRect(x, y, width, height int,
    ↪ rgba color.RGBA) {
    for idx := 0; idx < width; idx++ {
        for idy := 0; idy < height; idy++ {
            p.SetColor(x+idx, y+idy, rgba)
        }
```

```go
    }
}

func (p *Pixels) SetColor(x, y int, rgba color.RGBA) {
    r, g, b, a := rgba.RGBA()
    index := (y*p.Width + x) * 4
    p.Pix[index] = uint8(r)
    p.Pix[index+1] = uint8(g)
    p.Pix[index+2] = uint8(b)
    p.Pix[index+3] = uint8(a)
}

type GameOfLife struct {
    gameBoard [][]int
    pixels    *Pixels
    size      int
}

//Create a new GameOfLife structure with width*height number
//of cells.
//Size control how big to render the board game
func NewGameOfLife(width, height, size int) *GameOfLife {
    gameBoard := make([][]int, height)
    for idx := range gameBoard {
        gameBoard[idx] = make([]int, width)
    }
    pixels := NewPixels(width*size, height*size)
    return &GameOfLife{gameBoard: gameBoard, pixels: pixels,
        ↪ size: size}
}
func (gol *GameOfLife) Random() {
    for idy := range gol.gameBoard {
        for idx := range gol.gameBoard[idy] {
            gol.gameBoard[idy][idx] = rand.Intn(2)
        }
    }
}
```

```go
func CountNeighbors(matrix [][]int) [][]int {
    neighbors := make([][]int, len(matrix))
    for idx, val := range matrix {
        neighbors[idx] = make([]int, len(val))
    }
    for row := 0; row < len(matrix); row++ {
        for col := 0; col < len(matrix[row]); col++ {
            for rowMod := -1; rowMod < 2; rowMod++ {
                newRow := row + rowMod
                if newRow < 0 || newRow >= len(matrix) {
                    continue
                }
                for colMod := -1; colMod < 2; colMod++ {
                    if rowMod == 0 && colMod == 0 {
                        continue
                    }
                    newCol := col + colMod
                    if newCol < 0 || newCol >=
                                        len(matrix[row]) {
                        continue
                    }
                    neighbors[row][col] += matrix[newRow]
                        ↪ [newCol]
                }
            }
        }
    }
    return neighbors
}
func (gol *GameOfLife) PlayRound() {
    neighbors := CountNeighbors(gol.gameBoard)
    for idy := range gol.gameBoard {
        for idx, value := range gol.gameBoard[idy] {
            n := neighbors[idy][idx]
            if value == 1 && (n == 2 || n == 3) {
                continue
            } else if n == 3 {
```

```go
            gol.gameBoard[idy][idx] = 1
            gol.pixels.DrawRect(idx*gol.size, idy*gol.size,
                ↪ gol.size, gol.size, Black)
        } else {
            gol.gameBoard[idy][idx] = 0
            gol.pixels.DrawRect(idx*gol.size, idy*gol.size,
                ↪ gol.size, gol.size, White)
        }
    }
  }
}

func run() {
    size := float64(2)
    width := float64(400)
    height := float64(400)
    cfg := pixelgl.WindowConfig{
        Title:  "Conway's Game of Life",
        Bounds: pixel.R(0, 0, width*size, height*size),
        VSync:  true,
    }
    win, err := pixelgl.NewWindow(cfg)
    if err != nil {
        panic(err)
    }
    gol := NewGameOfLife(int(width), int(height), int(size))
    gol.Random()
    for !win.Closed() {
        gol.PlayRound()
        win.Canvas().SetPixels(gol.pixels.Pix)
        win.Update()
    }
}

func main() {
    pixelgl.Run(run)
}
```

When you run the program, you should see a window like Figure 4.6, where the Game of Life is chaotically killing and birthing (giving birth to) new cells!

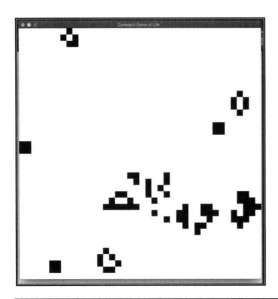

FIGURE 4.6   A snapshot of the Game of Life application in action

That's amazing! You've not only built a real application in Go, you've built an entire GUI for it as well! Now, we're sorry, but we want you to build another application *without* a user interface. Don't worry—it's still going to be fun—we'll implement a building block of one of the most hyped technologies in the world: Blockchain!

Proof of Work

The name and logic for this example app sound scarier than they actually are—the basic idea is actually pretty simple! We're essentially trying to brute force a value that's mathematically impossible to compute otherwise.

In fact, the code for this application is about 20 lines—it's not much, but you'll see why this is a specifically important example in the very next chapter, Chapter 5!

We must start this example with a bit more fundamental understanding. There exists a class of functions known as "cryptographic hashing functions." The idea behind these functions is that they can map an arbitrary sequence of bytes to a known size. A very small perturbation in the input can cause a drastic change in the output. And, because the output is of fixed size, and the input can be infinitely large, there is an infinite number of inputs that can map to a certain output.

When you have two separate inputs that map to the same output through a hashing function, it's known as a "hash collision." These can be issues for some systems. However, because of the sheer number of possible outputs, which we'll get into in a moment, the chances of this ever happening are so low that they're essentially negligible, so we basically ignore them.

One of the most popular hashing functions is "SHA256". It outputs a hash that's 256 bits (32 bytes) long. To give you an idea of what we mean by "a small change in the input can drastically affect the output," here's the hex representation of the SHA256 hash for "Tanmay Bakshi":

```
0x315920ffce53870d99f349ed8cebaab5c30f8efcc2f6ff835fe594e2539
↪ 28b54
```

Now, the following is the hash for "Tbnmay Bakshi". Only one character changed, and even that only from an "a" to a "b," just a single value.

```
0x6d3d44c306e9367484621e6639171226fc3b2daff31a8e60efb06168f2e
↪ 7904f
```

As you can see, it's not even remotely similar! It's essentially impossible to crack these hashes—it would take many decades worth of intense computing power—say, at a top-10 supercomputer level—just to make a dent in solving the problem, and by the time your descendants do figure out the solution, the data the hash is protecting will have lost its value.

However, there is a specific technology that found a very clever way to use this attribute of hashing functions: blockchain. Within blockchain, the goal is to verify the authenticity of a "block" of transactions in order to determine how much of a currency any specific wallet address has. We can verify a block's authenticity by determining how much "work" was put into trying to legitimize it.

For example, take the following string:

```
"Tanmay Bakshi + Baheer Kamal"
```

It's a pretty standard string, and the SHA256 hash for it looks like this in hex:

```
0x9D2BE96E68D9432AB2D883C51C2CD872DCACA6CA1495EF8A4E8865A9CCA
↪ 7FBA9
```

Now, imagine we wanted to append a certain sequence of bytes to the end of this string, such that the first few bytes of the output are all zeros. Of course, that would take an incredible amount of work, but it is possible. For example, when you find the hash for the following string:

```
"Tanmay Bakshi + Baheer KamalEHjjDodhxmxORke8hv0t"
```

You get this in hex format:

```
0x00000096408c134fb028dd260ce9e682e3524538442f34c874b4a4e7267
↪ 997d2
```

Wow! Just by appending that magic sequence, we were able to find a hash value that met our conditions! In fact, finding these many zeros took only around 13 million tries of random strings, and Go could do this in under 3 seconds, meaning it processed about 4 million random hashes per second.

Just to give you a bit more context, even though this doesn't really matter for this example, within Blockchain tech, you wouldn't have a string like "Tanmay Bakshi + Baheer Kamal". You would have a byte sequence that represented payments and other transactions, and by finding that magic sequence that fulfills the proof of work condition, and by having a reference to this block in the next block of transactions you create, you can make sure that people can't tamper with transactions that have already happened. That's because if they do tamper with a block, the whole chain of proof of work will fail to work. Unless, by some astronomically unlikely coincidence, they find an algorithm that returns exactly the right hashes that the blockchain needs—which is something we don't need to worry about.

The actual code that goes behind this fits snugly within a single function. Before we can get to it, though, we need to first implement a few utility and helper functions. Even before that, we need to get our imports out of the way:

```
import (
    "github.com/dustin/go-humanize"
    "time"
    "fmt"
    "crypto/sha256"
)
```

Now, helper functions. For example, we need a function that can help us generate random numbers:

```go
func RandomNumber(seed uint64) uint64 {
    seed ^= seed << 21
    seed ^= seed >> 35
    seed ^= seed << 4
    return seed
}
```

We know exactly what you're thinking: "Go's standard library doesn't even let us generate random numbers in peace!?" Well, yes, Go *does* already provide us functions that we can use to generate random numbers. However, the ones built into the standard library are really good; in fact, too good for us. What does that mean?

Remember, computers can't really generate random numbers, they're deterministic. Technically, everything in the universe is deterministic at a macro level, and things are only supposedly probabilistic at a quantum level. The closest we can get to random in the universe is chaotic, like the weather. But that's beside the point. Computers implement algorithms known as "pseudorandom number generators" that use chaotic math in order to generate numbers that are "uniform enough" in their randomness for our use cases.

These algorithms are pretty cheap in terms of compute time, but they vary in complexity. Some are very simple and give "good enough" distributions of numbers, and some are very expensive and provide cryptographically secure random numbers. For our use case, we don't need anything expensive, we only need super-simple random numbers. That's why, in this case, we implement our own very simple PRNG (pseudorandom number generator).

The reason we want it to be so cheap in terms of CPU cycles is because this function will be called many tens or hundreds of millions of times, so the CPU cycles on this function really add up to a lot of runtime if you're not careful. It just goes to show that when you're optimizing your code, you really need to optimize the hotspots, even if they seem simple. Complex code that's only ever called a couple of times doesn't need to be well-optimized.

Back to the function we implemented . . . The random number generator simply takes a random number seed, runs a few bit-level operations on it, and returns that new number. The idea behind its usage is that when we want a random number in a certain range, we'll feed a seed into this function, take the result, and run it through a modulus operation against the length of that range. Here's an example:

```
// generate random number from 0 to 9 (inclusive)
seed := 42 // this can be anything
seed = RandomNumber(seed) // this generates the new seed
rn := seed % 10 // this is your random number
seed = RandomNumber(seed) // this generates another new seed
rn := seed % 10 // this is another random number
```

Remember, when you start this chain of random numbers with the same seed, you will get the same random numbers; computers are deterministic, after all. This is why we start the seed with something that changes rapidly, such as a timestamp of the number of seconds since 1970.

The preceding code isn't going to be used exactly in our upcoming code, but that's what it's going to be based off of.

Next, in the same vein, we need a function that can generate random strings for us. This function is going to need a global variable that it can pull characters from to use in the random strings. Here's an implementation of both the global variable and the function:

```
var characterSet = []byte("abcdefghijklmnopqrstuvwxyzABCDEFGHIJ
↳ KLMNOPQRSTUVWXYZ1234567890")

func RandomString(str []byte, offset int, seed uint64) uint64 {
    for i := offset; i < len(str); i++ {
        seed = RandomNumber(seed)
        str[i] = characterSet[seed%62]
    }
    return seed
}
```

This function works in a pretty simple way. Before the function declaration, we define a new global variable called characterSet. This variable is an array of bytes, and this array is defined from a single string—each character essentially becomes its own element in this array. We will choose characters for the random string from this array.

The function itself takes a slice of bytes containing an entire string like "Tanmay Bakshi + Baheer Kamalxxxxxx" and an offset, telling the function where the "xxxxxx" part begins inside the string. We also take a random seed.

We then loop from the beginning of the random section (the offset) to the end of the string, at each index, placing a random character from the character-Set, choosing them with the PRNG that we coded a moment ago.

At the end, the function only needs to return the new seed value. The way this function is designed makes it so we don't need to make copies of the string in memory whenever we want to try out a new hash, we can keep reusing the same space in memory with the next set of random characters changed for each iteration. This makes the code even faster.

Finally, let's implement a helper function that can take a string and a proof of work condition, and tell us if we've met that condition:

```go
func Hash(data []byte, bits int) bool {
    bs := sha256.Sum256(data)
    nbytes := bits / 8
    nbits := bits % 8
    idx := 0
    for ; idx < nbytes; idx++ {
        if bs[idx] > 0 {
            return false
        }
    }
    return (bs[idx] >> (8 - nbits)) == 0
}
```

This function takes both the data it needs to check against, as well as the number of consecutive starting bits in the hash we expect to be all zeros. It does this in a few simple steps:

1. Calculate the SHA256 hash of the data using the built-in crypto package from Golang.
2. Calculate the number of bytes that we need to be all zero.
3. Calculate the number of bits after the bytes that we need to be all zero.
4. Loop through the bytes and check them at a high level, returning "false" if any byte has a non-zero value (we didn't meet the proof of the work condition).
5. If the bytes were fine, we run a bitshift to check if the final few bits also meet the PoW condition. If they do, we return true; otherwise, we return false.

We're done with our helper functions now! Let's finally move on to the main part of this code, the function that actually finds a matching sequence of random characters:

```go
func pow(prefix string, bitLength int) {
    start := time.Now()

    totalHashesProcessed := 0
    seed := uint64(time.Now().Local().UnixNano())
    randomBytes := make([]byte, 20)
    randomBytes = append([]byte(prefix), randomBytes...)
    for {
        totalHashesProcessed++
        seed = RandomString(randomBytes, len(prefix), seed)
        if Hash(randomBytes, bitLength) {
            fmt.Println(string(randomBytes))
            break
        }
    }

    end := time.Now()

    fmt.Println("time:", end.Sub(start).Seconds())
    fmt.Println("processed",
        ↪ humanize.Comma(int64(totalHashesProcessed)))
    fmt.Printf("processed/sec: %s\n", humanize.Comma(int64(
        ↪ float64(totalHashesProcessed)/end.Sub(start)
        ↪.Seconds())))
}
```

As you can tell, this function can remain pretty simple thanks to the helper functions that we've created. The function begins by noting the time at which it started execution. Using this value, at the end of the function, we can calculate the amount of time it took for us to find a solution, and the number of hashes we could process per second.

We then create a couple of variables that we'll need, such as the total number of hashes that have been checked, the random seed (which is based off of a timestamp), and the byte array representing the prefix string with a bunch of random characters concatenated to it.

Next, we run an infinite loop in which we always start by incrementing the number of hashes processed, because we know we're going to process a hash if we've reached this stage of execution. Then, we fill in new random characters in our string and refresh our random seed. We check if we've met the proof of work condition, and if we have, we print the solution and break the loop. If not, we continue the loop iterations until we hit that break.

Then, we get the end time, and print out some statistics. We use a module known as "humanize" in order to place commas in our large numbers so they're more readable.

There's just one more thing we need to do: add a main function that calls pow! It's quite simple to do so:

```
func main() {
    pow("Tanmay Bakshi + Baheer Kamal", 24)
}
```

Your code should look like this:

CODE LISTING 4.3   "Proof of work" application

```
package main

import (
    "github.com/dustin/go-humanize"
    "time"
    "fmt"
    "crypto/sha256"
)

var characterSet = []byte("abcdefghijklmnopqrstuvwxyzABCDEFGHIJ
↳ KLMNOPQRSTUVWXYZ1234567890")

func RandomNumber(seed uint64) uint64 {
    seed ^= seed << 21
    seed ^= seed >> 35
    seed ^= seed << 4
```

```go
        return seed
    }


    func RandomString(str []byte, offset int, seed uint64) uint64 {
        for i := offset; i < len(str); i++ {
            seed = RandomNumber(seed)
            str[i] = characterSet[seed%62]
        }
        return seed
    }


    func Hash(data []byte, bits int) bool {
        bs := sha256.Sum256(data)
        nbytes := bits / 8
        nbits := bits % 8
        idx := 0
        for ; idx < nbytes; idx++ {
            if bs[idx] > 0 {
                return false
            }
        }
        return (bs[idx] >> (8 - nbits)) == 0
    }


    func pow(prefix string, bitLength int) {
        start := time.Now()

        totalHashesProcessed := 0
        seed := uint64(time.Now().Local().UnixNano())
        randomBytes := make([]byte, 20)
        randomBytes = append([]byte(prefix), randomBytes...)
        for {
            totalHashesProcessed++
            seed = RandomString(randomBytes, len(prefix), seed)
            if Hash(randomBytes, bitLength) {
                fmt.Println(string(randomBytes))
                break
```

```
                }
        }
        end := time.Now()

        fmt.Println("time:", end.Sub(start).Seconds())
        fmt.Println("processed",
            ↪ humanize.Comma(int64(totalHashesProcessed)))
        fmt.Printf("processed/sec: %s\n", humanize.Comma(int64
            ↪ (float64(totalHashesProcessed)/end.Sub(start)
            ↪ .Seconds())))
}


func main() {
    pow("Tanmay Bakshi + Baheer Kamal", 24)
}
```

In this case, we're asking the function to find 24 consecutive starting bits of zeros. You can replace the string with any string you'd like, and you can also tell it to find as many zeros as you want to! Remember, though, that as you get larger and larger numbers of how many bits of zeros you'd like the hash to start with, the task gets harder and harder—not linearly, so every extra bit adds more difficulty than the one added before it.

If you run this code as is, because it's quite optimized to begin with, you'd get output that looks something like this:

```
Tanmay Bakshi + Baheer KamalEHjjDodhxmxORke8hv0t
time: 2.7946521730000002
processed 12,814,971
processed/sec: 4,585,533
```

It might seem impressive to get 4.5 million hashes per second, but in the computing world, these are rookie numbers. We should be impressed when we get 30 million per second on CPU. Or maybe 8 *billion* per second on GPU! The sky is the limit when it comes to making this code faster.

One immediate potential speedup is using *concurrency*, so we have multiple processes looking at random strings at the same time. However, you'll have to read Chapter 5 to learn about that!

Now that you've learned how to implement some common algorithms and techniques in Go, let's level up this code! We'll use a language feature that really sets Go apart, Goroutines, and learn how to optimize our code to make it run faster! Every time you break a new optimization and speed barrier, you end up learning enough new stuff to feel like you've become a better programmer!

Exercises

1. What are some instances in which we went out of our way in this chapter's examples to optimize the code so it runs faster?

2. How can Dijkstra's algorithm, meant to work on graphs, be repurposed to work on a "map," such as a maze?

3. Which operator is meant for bitwise: NOT, OR, AND, or XOR? Write a program to test the behavior of these operators.

4. Can computers generate truly random numbers? Why or why not?

5. What kind of expensive operation can be avoided by not copying variables?

6. In Code Snippet 4.1 for Dijkstra's pathfinding algorithm, try changing the cost from B to D from 1 to 10. What do you predict the new output will be, and does it match what the program computes?

7. What's the likelihood of any two pieces of data encountering a hash collision with SHA256?

Concurrency

W elcome to Chapter 5! You've progressed quite far down your journey of learning Go, and you're now at the level where you can competently implement simple and intermediately complex applications in Go. However, to really master Go, and to unlock its full potential, you need to be familiar with a cornerstone in modern computing: concurrency. By enabling a computer to do multiple things at the same time—or by switching across multiple execution streams—you can run tasks that computers of the past could never have done.

Once you've finished this chapter, you'll be able to answer the following questions:

- What is a thread and a process?
- What's the difference between concurrency and parallelism?
- How do green threads and user-space threads work?
- Why are Goroutines better than using threads? When should you *not* use them?
- How can you communicate among different Goroutines with channels?
- How can a Goroutine wait for input from multiple sources at once?

Concurrency, Threads, and Parallelism

Concurrency is fundamental to modern computing; it enables computers to execute independent streams of instructions simultaneously, thereby enabling functionality such as multiple applications running at the same time.

Imagine if your computer could only do a *single thing* at once. It would be incredibly limiting! We wouldn't even have the technical provision for graphical interfaces. We'd instead be stuck with command-line interfaces for everything.

On some systems, concurrency simply means that the CPU will switch from running instructions of one process to another in very quick succession, making it look to us humans as if multiple things are happening at the same time.

However, this technique has its limitations. If you're trying to run multiple processes that require a lot of compute time, and if they all need to run at once, then the CPU will spend a *lot of time* running those processes. That's because, not only does the CPU need to do the computing per process, it also needs to keep switching between the processes, which in itself takes a bit of computing power.

This is why modern CPUs have "cores." The more cores a CPU has, the more independent streams of execution it can perform essentially simultaneously. This means that as the instructions of one process execute on one "core," a totally different process can be running its instructions in another "core" during the same "clock cycle."

Now, the real value of this technique starts to show when you merge the old and the new. Imagine having multiple physical cores, capable of running multiple different processes at the same time, and also having them switch between processes really fast so we can build systems such as GUIs which, by design, require concurrency! This is how modern computing systems are architected—all the way from the CPU to the operating system, and even to the programming languages.

We now know that a single CPU and OS instance can run multiple—meaning thousands—of different processes at the same time. However, what if an independent process needs to run its own multiple streams of execution at the same time? For example, let's say you have a web browser, and you open multiple tabs. Technically, the browser is a single process, but there are multiple tabs all running their own code. This is where *threads* come into play. A thread is a lightweight part of a process that has its own "stack pointer" and registers. A process could spawn hundreds of threads, and the operating system would map those threads to the CPU just as it would other processes.

Something to note: The browser analogy we gave isn't set in stone and there are a few notable exceptions. For example, on the Google Chrome web browser, each tab gets its own full-blown process, which makes it easier for Chrome to hit performance benchmarks, but it's very resource intensive. That's why Chrome is notorious for consuming orders of magnitude more battery power on laptops, as well as RAM, than other browsers such as Safari and Firefox that so far use threads.

You may now be thinking: "This all seems too good to be true! We can just run thousands of threads and processes and our operating system handles it all for us efficiently? Wow!" Unfortunately, when something seems too good to be true, it usually is. There are quite intense limitations to scaling with threads.

You see, threads may be relatively lightweight compared to full-blown processes, but they're still very expensive to spin up and then join with. It requires an entire system call, and interaction with the kernel, to do so. This means that in instances when you need thousands of threads, it's usually more efficient to only spin up tens or hundreds, and then have each thread do multiple pieces of work.

That's why, for most software today, there are actually two kinds of threads: native threads, which are the kind we've already talked about (handled at the kernel-level), and green threads, which we'll discuss in a moment (they're handled at the user-level).

Think about web servers for a moment. A single system may get thousands, maybe tens of thousands, of requests at once that it needs to deal with. In this case, threads would be insufficient, because there would be so many of them that the scheduler, which is responsible for switching between threads, would execute for a longer period of time than the threads would themselves.

That's why the web servers implement "green threads," which are handled by the server itself, making it so that, within a single thread, we can switch between multiple different green threads at certain points in the code. Because we have more fine-grained control over where a "context switch" happens, performance is usually enhanced for web server–type applications.

Go has a unique implementation of a similar concept, that follows a paradigm known as the "M:N" model of threading. This is also known as "hybrid threading." The M represents the user-mode or "green" threads, and the N represents the kernel-mode or "native" threads. It's similar to what you'd get if you were to combine completely kernel mode (1:1) and completely user mode (N:1) threading.

The extra complexity in hybrid threading comes with the fact that you need to modify both kernel- and user-level code to get the system to work. However, we do have something working in our favor: Most modern kernels have all the provisions we need already built in! We just need good user-level software to handle the green threads, and that's exactly what Go delivers on.

Goroutines

Go has a concept of "Goroutines" that map lots of different execution streams, which we can just call functions, to a pool of native threads. To give you an idea how lightweight Goroutines are, they require only 2KB of memory, compared to a native thread's 1MB plus a guard page, which is *500×* as much!

Before we dive into how these Goroutines work, let's take a look at an example of how you can use them.

The syntax for creating Goroutines is quite simple: To invoke a function as you normally do, you only need to add the token "go" before the invocation. For example, take a look at the following code:

CODE LISTING 5.1   A naïve, single-threaded program to square numbers

```go
package main

import (
    "fmt"
)

func squareIt(x int) {
    fmt.Println(x * x)
}

func main() {
    squareIt(2)
}
```

As you can imagine, when you compile and run this code, it prints out the number 4. But what if `squareIt` was really a long-running operation that you wanted to run in a separate Goroutine, so the main function can continue doing whatever it needs to do while `squareIt` runs? All you need to do is put "go" before `squareIt` in the main function:

```
func main() {
    go squareIt(2)
}
```

Great! If you compile your program, you'll notice there are no errors. However, when you run the code, you'll see that it prints . . . nothing. (This is usually the case on most systems.) Why is that?

(Even if it did print something out, continue reading. You'll see why in a moment.)

Well, think about it. When we ran the Goroutine, we put the execution of `squareIt` into the "background," which is another way of saying it's "a separate Goroutine from the one we're currently on." Then, right as we did that, the main function ended. The `main` function is executing in the main Goroutine, which is executing the main thread of the process this program is running in. When the `main` function ends, the entire program exits—and that means the process exits before our other Goroutine even had the chance to print out the squared number.

That means we need to tell our main thread that it needs to *wait* before exiting, so the other Goroutine has a chance to finish. This can be done quite simply by just telling the main function to "sleep" (idle) for, say, just a single millisecond. To do so, you first need to import the time package. Modify your imports to the following:

```
import (
    "fmt"
    "time"
)
```

And then add a call to the Sleep function within your main function:

```
func main() {
    go squareIt(2)
    time.Sleep(1 * time.Millisecond)
}
```

Now, if you compile and run your program, you should notice it prints out 4 like usual! If it does not, try to increase the time you sleep; your processor just might need a bit more time to compute.

Again, you should keep one thing in mind: This is not how we actually wait for Goroutines to finish in real-world scenarios. You should really be using a feature known as *channels*, which we'll get into in a bit.

The idea behind Goroutines is almost genius in terms of efficiency. When a program starts, Go only has a single Goroutine running, so there's only a single native thread running as well. Whenever you have multiple Goroutines running simultaneously, Go will map each Goroutine to multiple threads. If it needs to launch more threads to try and run Goroutines in parallel instead of just concurrently, it may decide to do that. In this case, that means it will launch another thread when you run `go squareIt(2)`.

So, for example, just like how the OS is able to map multiple threads to CPU cores, the Go runtime is able to map multiple Goroutines to native threads. And, by default, native threads can "move around" between different CPU cores (e.g., core 1 was executing thread 1 a few moments ago, but after a few context switches, core 2 is now executing thread 1). This can be locked down with certain flags, telling the kernel to only run a thread on a specific core. Similarly, you can tell Go "keep this Goroutine pinned to a specific thread." Again, take this with a grain of salt, given this is a kernel feature—most kernels (Linux, BSD, XNU, etc.) support it, but some may not.

Now here's the really interesting part: Let's say that while Go is multiplexing Goroutines on threads, and while it's handling scheduling, a certain Goroutine blocks these because it needs to execute a syscall (system call). These are usually the slowest parts of your program. In this case, Go will *create a new thread* and move Goroutines to that thread. This means that when executing a long-running operation that we don't need to idle the CPU to wait for, Go can keep other Goroutines running. Plus, that operation only requires storing three registers, meaning it's *super-lightweight* and your programs can run *really fast*! When you compare that to the cost of switching between threads, which require storing *every single register* (50+ registers on x86-64), you can see why Goroutines are valuable.

There is one disadvantage to the very lightweight nature of Goroutines, which Go has implicitly introduced a solution to. The problem is that when you spawn a new Goroutine, you no longer have any control over it from other Goroutines. For example, one Goroutine cannot "kill" another one or "merge" with another one. Instead, a Goroutine can only "finish executing" or "exit" once it returns (reaches the end of a function).

This limitation also means that the function you invoke as a Goroutine cannot return any values. You must instead use channels to share information.

Speaking of which, you now have a good idea of how the world of Goroutines works. Now let's take a look at another integral component of using Goroutines for concurrency: channels.

Channels

With channels, you have the ability to send data to different Goroutines in a very high-performance manner. Plus, their syntax is very simple to use, just as with Goroutines, resulting in code that doesn't get more complex because of communication between Goroutines.

Let's expand our previous number squaring example to use channels for communication. Let's update the squareIt function to take two arguments, both being channels. One channel is to let this function know what numbers it needs to square, and the other channel is where the Goroutine will put the numbers once it has squared them.

Here's the code for the squareIt function:

```
func squareIt(inputChan, outputChan chan int) {
    for x := range inputChan {
        outputChan <- x * x
    }
}
```

Before we take a look at the main function, let's decipher the squareIt function a bit more. As you can see, the type annotation for channels is unique in the sense that it is comprised of *two tokens* (chan int) whereas most type annotations are just a single token, int. However, the meaning of the annotation isn't ambiguous—we're telling Go that we want a channel (chan), and the data within that channel is of type int. You could store any data type, including pointers and custom structures, within a channel.

As we mentioned, this function takes two separate channels: an input channel and an output channel. Within the function, we loop through the input channel just like how we loop through essentially any other sequence, like a dictionary or array. However, what Go does behind the scenes is a little bit different. When there are no values in the channel, the for loop will block and start waiting for a new data point. When it finds one, then the `for` loop continues with its iteration, and at the end, jumps back to the top of the loop.

Within the `for` loop itself, we do only one simple thing: We square "x", which is the integer we got from the input channel, and then we put that squared value into the output channel using the "<-" operator.

As you can tell, modifying this function to use channels was quite simple. However, the `main` function still requires a bit more rearchitecting:

```go
func main() {
    inputChannel := make(chan int)
    outputChannel := make(chan int)
    go squareIt(inputChannel, outputChannel)
    for i := 0; i < 10; i++ {
        inputChannel <- i
    }
    for i := range outputChannel {
        fmt.Println(i)
    }
}
```

Let's go through the `main` function section by section again.

We start with the first two lines of code that are responsible for creating the input and output `channels` using the "`make`" function, which we used in Chapter 2 to create arrays:

```go
inputChannel := make(chan int)
outputChannel := make(chan int)
```

Then, we run the `squareIt` function in a separate Goroutine:

```go
go squareIt(inputChannel, outputChannel)
```

At this point, the function is running in the background, and because of the `for` loop, it's waiting for data to come through the input channel.

In order to feed the input channel with data, we simply loop from 0 to 9, and at each iteration we place the number within the input channel:

```go
for i := 0; i < 10; i++ {
    inputChannel <- i
}
```

At this point, because we've put the data into the channel, we expect that the squareIt function has picked up the data and placed it in the output channel. Then, we need to loop through the output channel to get the data back from the squareIt function and print it out so we can see that output too:

```
for i := range outputChannel {
    fmt.Println(i)
}
```

If you've been really attentive up to this point, you may have noticed a bug in this program. And, if you already know about how channels work, you've likely noticed *another* bug in this program. However, let's roll with this code for now, and determine what the bugs are when we get to them.

Here is the final code:

CODE LISTING 5.2   A concurrent implementation of the number squaring application

```
package main

import (
    "fmt"
)

func squareIt(inputChan, outputChan chan int) {
    for x := range inputChan {
        outputChan <- x * x
    }
}

func main() {
    inputChannel := make(chan int)
    outputChannel := make(chan int)
    go squareIt(inputChannel, outputChannel)
    for i := 0; i < 10; i++ {
        inputChannel <- i
    }
    for i := range outputChannel {
        fmt.Println(i)
```

```
        }
    }
```

First, compile the program. Notice that it compiles with no errors. However, when you run the code, on the majority of systems you'll see an error like this one:

```
fatal error: all goroutines are asleep - deadlock!

goroutine 1 [chan send]:
main.main()
    /tmp/sandbox306189119/prog.go:18 +0xb4

goroutine 6 [chan send]:
main.squareIt(0xc00005e060, 0xc00005e0c0)
    /tmp/sandbox306189119/prog.go:9 +0x4e
created by main.main
    /tmp/sandbox306189119/prog.go:16 +0x8e
```

All right, so we know there *is* an issue with the logic of our program. In fact, Go gives us a pretty clear error message here; we've encountered what's known as a *deadlock*. A deadlock occurs when all of your threads are waiting on another one to do something, so no thread gets the opportunity to do the thing that the others are waiting for.

In most programming languages, especially the ones that use native threads, this just results in your program freezing and refusing to cooperate, and at this point you must kill the process. However, because the Go runtime can see what each Goroutine is doing, it sees the deadlock and kills the process for you. Plus, it prints a neat little stack trace to help you determine where your code is failing!

Currently, the deadlock arises because we're using what are known as *unbuffered* channels. An unbuffered channel means the channel has no "storage space," and when a sender puts some data within the channel, a receiver needs to take that data immediately. If a receiver has not taken the data, then, as you try to put more data into the channel, you will be blocked until the channel is empty, at which point you'd be allowed to place the data inside it.

So, Table 5.1 is a timeline of what's happening through the execution of the program.

TABLE 5.1   A Walkthrough of How Both Goroutines' Execution Leads to a Deadlock

MAIN GOROUTINE	SQUAREIT GOROUTINE
Create the input and output channels	[nil]
Spawn the `squareIt` Goroutine	Wait for data on the input channel
Place the number 0 in the input channel	
Wait for the number 0 to be taken from the input channel	Grab the number 0, square it, place it in the output channel
Place the number 1 in the input channel	
Wait for the number 1 to be taken from the input channel	Grab the number 1, square it, wait for the number 0 to be taken from the output channel
Place the number 2 in the input channel	
Wait for the number 2 to be taken from the input channel	[still waiting for 0 to be taken from the output channel]
[still waiting for 2 to be taken from the input channel]	

At this point, both Goroutines are waiting for each other to do something that they both won't get the opportunity to do because . . . they're waiting for each other. In order for either of these Goroutines to continue executing, the other one would need to execute as well. As you can likely tell, this deadlock wouldn't occur if the `squareIt` Goroutine didn't need to wait for another Goroutine to take the results out of the output channel.

In order to solve this problem, we can use what's known as a "buffered" channel. In a buffered channel, you can insert up to only a certain number of elements that you define asynchronously (without blocking) before the channel blocks and waits for a Goroutine to take data out before it lets you add more. Essentially, the channel lets you "buffer" or "store" a certain number of elements without blocking.

You can create a buffered channel by simply passing the size of the buffer to the `make` function when you're making the channel. For example, modify the output channel instantiation to the following:

```
outputChannel := make(chan int, 10)
```

And now, if you compile your code, it should still work. Plus, when you run your code, success! You see the numbers you expect!

```
0
1
4
9
16
25
36
49
64
81
fatal error: all goroutines are asleep - deadlock!

goroutine 1 [chan receive]:
main.main()
    /tmp/sandbox997165491/prog.go:20 +0x157

goroutine 6 [chan receive]:
main.squareIt(0xc000066000, 0xc000068000)
    /tmp/sandbox997165491/prog.go:8 +0x66
created by main.main
    /tmp/sandbox997165491/prog.go:16 +0x8e
```

Well, at least it was *kind of* a success. If you see where it deadlocks, line 16 of the code (as pointed out by Go in the last line of the stack trace), you'll realize the bug now arises with the way we take data out of the output channel.

See, when the output channel is empty, and the input channel is empty, then the main function is blocked because it's waiting for data on the output channel, and there's nothing that'll come in the output channel because the input channel is empty and the other goroutine is waiting for data on the input channel.

The way we can fix this is by replacing this loop in the main function:

```
for i := range outputChannel {
    fmt.Println(i)
}
```

This loop will block when there's nothing in the output channel because the loop does not know when to stop. However, in this case, we know there are only going to be 10 values, so we can replace the for-in loop with this:

```
for i := 0; i < 10; i++ {
    fmt.Println(<- outputChannel)
}
```

Now, when we've done 10 iterations, the channel will still technically be open, meaning we *could* be getting more data and we wouldn't know it, but because we know the logic of the rest of the program, we know this won't happen.

Also, as you can see, we also use the "<-" operator to take data out of a channel.

Before we move on to the next major pillar of Go's concurrency, we think it's helpful to visualize the way a channel works in the backend.

We've already seen two kinds of channels:

- **Unbuffered Channels (sync channels).** These need to store a single element from a single Goroutine, wait for some Goroutine to pick it up, and then it's available to store another element.
- **Buffered Channels (async channels).** These need to store multiple elements from potentially multiple Goroutines, and only block the senders when the channel is full. It needs to allow potentially multiple Goroutines to receive elements as well.

However, there is also a third kind of channel that we haven't taken a look at yet: the channel with zero-sized elements. For example, imagine a channel like this:

```
channel := make(chan struct{})
```

In this case, the type of the data in the channel is just an empty structure. When this is the case, we're telling Go "I'm not storing anything in the channel, I'm just using this to help me synchronize my threads." In other words, you could look at this as like a semaphore.

So, for example, take a look at this Swift code:

```
let a = DispatchSemaphore(value: 0)
DispatchQueue.global(qos: .background).async {
    a.signal()
    print("signalled")
}
a.wait()
print("exiting")
```

And take a look at this C code:

```c
#include <stdio.h>
#include <pthread.h>
#include <semaphore.h>
#include <unistd.h>

sem_t semaphore;

void *backgroundThread(void *vargp) {
    sem_post(&semaphore); // Signal to the semaphore
    printf("signalled\n"); // Print that we signalled
    return NULL;
}

int main() {
    sem_init(&semaphore, 0, 1); // Initialize the semaphore
    // Launch a new thread
    pthread_t thread;
    pthread_create(&thread, NULL, backgroundThread, NULL);
    sem_wait(&semaphore); // Wait for a signal on the semaphore
    printf("exiting\n"); // Print that we are exiting
    sem_destroy(&semaphore); // Destroy the semaphore
    pthread_join(thread, NULL); // Join with the thread we
                                // launched
    return 0;
}
```

The logic for both of these samples should be mostly self-explanatory. You create a semaphore, and then launch a new thread that does some task and then releases the semaphore once it's done. Then, while that's going on in the background, the main thread starts to wait on that semaphore, essentially waiting for the background thread to be done with its task.

In Go, you can replicate this functionality by using a channel with zero-sized elements. Here's an example:

```go
func main() {
    semaphore := make(chan struct{})
    go func() {
```

```
        semaphore <- struct{}{}
        fmt.Println("signalling")
    }()
    <- semaphore
    fmt.Println("exiting")
}
```

Because this is technically an unbuffered channel (sync channel), only one "signal" can be in the semaphore at once. In order to support more than one signal, you can simply make it a buffered channel (async channel):

```
semaphore := make(chan struct{}, 10)
```

The way these channels are implemented is by using a circular linked list data structure. Each node in the linked list contains information about the Goroutine that sent the data, as well as a pointer to the data itself.

Every Goroutine that's waiting for data is making "laps" around the circular Goroutine, racing the other Goroutines to find a piece of data from a sender within the linked list. When a Goroutine finds some data, it attempts to lock it, and if it wins the race, it gets to use that data and mark it as "stale" within the linked list, so another Goroutine doesn't pick up the same data.

Remember, there are many kinds of channels, and this describes the basic working of the async kind of channel. However, other channels have similar internals with some small differences.

However, there is just one more part of this program that is "inelegant." It's that when the main Goroutine exits, it kills the `squareIt` Goroutine, but that Goroutine never exits gracefully. That's because all it's doing is waiting for data from the input channel, and it's never looking to see if it's done with its work.

There are two ways for us to solve this problem in this specific case. One way is to use a built-in feature of channels, and the other is to introduce the next major pillar in Go's concurrency design, which you will learn about in the next section. Let's start off by using features built into the channels we already know and love.

In this case, we can solve our problem by "closing" the channel. This makes it so you can no longer send data into the channel, and any Goroutine that's currently blocked waiting for the channel will be unblocked.

All you need to do is add this line of code to the very end of your main function:

```
close(inputChannel)
```

Now, your main function should look like this:

```go
func main() {
    inputChannel := make(chan int)
    outputChannel := make(chan int, 10)
    go squareIt(inputChannel, outputChannel)
    for i := 0; i < 10; i++ {
        fmt.Println(i)
        inputChannel <- i
    }
    for i := 0; i < 10; i++ {
        fmt.Println(<- outputChannel)
    }
    close(inputChannel)
}
```

When you run it, you should see the output you expect, because we've fixed the deadlocks:

```
0
1
2
3
4
5
6
7
8
9
0
1
4
9
16
25
36
49
64
81
```

While the output doesn't look different, you can rest peacefully knowing your Goroutines are exiting gracefully. In this code, that doesn't really matter, because it would've been killed as your program exited anyway, but this can be useful when you know you need certain Goroutines to exit.

select Statements

There is, however, another way to solve this problem, and also introduce new functionality to channels: `select` statements.

Think of `select` as "switch, but for channels." With `select`, you're allowed to have a Goroutine wait for multiple channels, and then take data from whichever one provides a value first. Imagine we wanted to create a function that not only does squaring, but also does cubing and needs to exit gracefully.

In that case, we could "simply" implement a function like this:

```
func squarerCuber(sqInChan, sqOutChan, cuInChan, cuOutChan,
    ↪ exitChan chan int) {
    var squareX int
    var cubeX int
    for {
        select {
        case squareX = <- sqInChan:
            sqOutChan <- squareX * squareX
        case cubeX = <- cuInChan:
            cuOutChan <- cubeX * cubeX * cubeX
        case <- exitChan:
            return
        }
    }
}
```

This function may look complex, but when you break it down into its components, it's actually pretty simple.

The basic logic of the function is that it gets numbers to square and cube from the `sqInChan` and `cuInChan` channels, respectively. Then, once it does the operation it's supposed to, it puts the results into `sqOutChan` or `cuOutChan`. Finally, when the function gets any message from `exitChan`, it knows it should return (exit).

The code for the function starts off by just telling Go "we will have two variables of type integer in the future, called `squareX` and `cubeX`." These variables will contain the integers we get from the input channels.

Then, we run an infinite `for` loop. Within this `for` loop, we have the new player in the arena: the `select` statement. As you can see, it looks a lot like a `switch` statement with its `cases`. However, the `cases` are a bit different this time. Every `case` is an expression that can only be resolved when a blocking communication operation unblocks and returns a response. For example, take a look at the first `case`:

```
case squareX = <- sqInChan:
```

Here, we're telling Go "Hey! Here's a new `case`: We want to place an integer value inside of `squareX` that comes from `sqInChan`." However, because this operation may block if there are no values inside of the `sqInChan` channel, this expression cannot be resolved until the channel's "receive" operation unblocks.

This is where `select` comes in. `select` will look at all of these `cases`, where the expressions are being blocked by something, and whichever one unblocks first will get to run its code. After that code runs, control flow goes through the statement to whatever code is next. This means it's not a loop; it only runs a single section of code. This is why we need that infinite `for` loop to wrap the `select` statement.

In order to actually use this function as a background Goroutine, you need a `main` function to drive it and get the output from it. Here's an example of implementing a `main` function which does exactly that:

```
func main() {
    sqInChan := make(chan int, 10)
    cuInChan := make(chan int, 10)
    sqOutChan := make(chan int, 10)
    cuOutChan := make(chan int, 10)
    exitChan := make(chan int)
    go squarerCuber(sqInChan, sqOutChan, cuInChan, cuOutChan,
    ↳ exitChan)
    for i := 0; i < 10; i++ {
        sqInChan <- i
        cuInChan <- i
    }
```

```
    for i := 0; i < 10; i++ {
        fmt.Printf("squarer says %d\n", <- sqOutChan)
        fmt.Printf("cuber says %d\n", <- cuOutChan)
    }
    exitChan <- 0
}
```

The working of this `main` function is very similar to the previous one. The only real difference is that we have two input and two output channels, and we have a new channel for telling the Goroutine to exit. Plus, within the `for` loops that are responsible for sending data to and receiving data from the channels, we do so for *both* the squaring and cubing channels.

The last difference is that, at the end, we insert a dummy value into the exit channel, in order to tell the Goroutine that its work is done, and that it can exit.

Also, be sure to keep in mind that with Go, implementing the exit functionality is fundamentally your responsibility! In this case, the exit channel only tells the Goroutine to exit because we coded it that way. However, it's always possible to do something totally different in the background other than exiting. With OS threads, you do at least always have the option to kill or join the thread, but with Goroutines, you take responsibility of making sure they know when to shut themselves down.

Before we continue, there is one more point we want to cover: When you invoke a Goroutine, you technically are making a function invocation, and telling Go to actually run that invocation in a separate Goroutine. However, so far, we've only shown you how to invoke functions that you've already defined. What if you want to create and invoke a function in a separate Goroutine at the same time? Well, Go lets you do just that:

CODE LISTING 5.3   Invoking a function in a separate Goroutine in-place, as you define it

```
func main() {
    inputChan := make(chan int, 10)
    finishChan := make(chan int)
    outputChan := make(chan int, 10)
    go func(inputChan, finishChan chan int) {
        for {
            select {
            case x := <- inputChan:
```

```
                outputChan <- x * x
            case _ = <- finishChan:
                return
            }
        }
    }(inputChan, finishChan)
    for i := 0; i < 10; i++ {
        inputChan <- i
    }
    for i := 0; i < 10; i++ {
        fmt.Println(<- outputChan)
    }
    finishChan <- 1
}
```

In this case, the logic is quite simple, but the syntax may be foreign to you. The function signature and body are embedded directly in the callsite. This is valid Go syntax. Line 5 of the snippet tells Go "I want you to launch another Goroutine, and here's the function you're going to run." The rest is like a regular function signature, but without the function name. Then, on line 14, you tell Go "all right, that was the function; now I want you to call the function and pass `inputChan` and `finishChan` to it as arguments."

When you call Goroutines, remember that Go differs from the more common way of dealing with variables and scope. For example, try running this code:

CODE LISTING 5.4   Demonstrating Go's variable capture for closures

```
func main() {
    for i := 0; i < 10; i++ {
        go func() {
            time.Sleep(1 * time.Millisecond)
            fmt.Println(i)
        }()
    }
    time.Sleep(100 * time.Millisecond)
}
```

In most programming languages, and even just by intuition, this code should print the numbers 0, 1, 2, 3, 4, 5, 6, 7, 8, 9 in some random order, because concurrency implemented in this specific way isn't deterministic. However, in Go, the compiler will first throw a warning like so:

```
loop variable i captured by func literal
```

And then, it'll print:

```
10
10
10
10
10
10
10
10
10
10
```

Ten tens! Why is that? It's because the function we call within the loop is doing what's known as "capturing" the loop variable. In most languages, when a "closure" or "function literal" tries to capture a variable, it's given its own copy. In object-oriented languages, this works especially well because, if we're talking about a class object, we're only copying the reference to the aforementioned object.

However, Go takes a different path. It gives us access to the loop variable itself, instead of making a copy. So, in this case, we're going through the loop and spawning a bunch of Goroutines, each of which starts sleeping. At the end of the loop, once we increment i to 10, the for loop says "oh we're done iterating now" and stops spawning Goroutines.

Around this time, all the Goroutines wake up, and because they're all referring to the same "i", they all print 10, instead of the value of i at the time they were called.

This problem can be solved pretty easily by actually passing the loop variable to the function instead of trying to capture it. This way, Go needs to call the function and pass the value as a parameter, forcing it to make a copy:

CODE LISTING 5.5   Forcing Go to make a copy of a variable to pass to a closure

```go
func main() {
    for i := 0; i < 10; i++ {
        go func(nonCapturedI int) {
            time.Sleep(1 * time.Millisecond)
            fmt.Println(nonCapturedI)
        }(i)
    }
    time.Sleep(100 * time.Millisecond)
}
```

If you were to run this code, you would see it outputs what we expect: numbers 0 through 9 in random order.

All right, we can now say with confidence that you have gained enough experience with Goroutines to implement a more complex example of this technology! Specifically, it's time to take an application we've already built in Chapter 4 and upgrade it using the power of concurrency.

Proof of Work: Part 2!

If you recall the last example in Chapter 4, you'll remember we implemented a simple version of a "proof of work" algorithm, one of the key components of a blockchain. However, we only reached around 4 million hashes per second being processed by our program, and that was after a pretty well optimized implementation. So how can we make this better?

The answer is *concurrency*! This task in particular is very well-suited for applying concurrency. You only need multiple instances of your program running at once that are all looking for a solution. These instances don't really need to talk to each other, making it so we don't need extra overhead just because we decide to use concurrency.

Remember that making your code concurrent is definitely *not* always a free ticket to extra performance. There is overhead associated with everything; you're never just going to get free lunch. There is overhead in spawning a Goroutine or a thread, in context switching, in communicating, in spinning it down, and the extra safety checks that come with all of this. Plus, it's very easy to accidentally end up implementing pieces of functionality that hit problems like Cache

Invalidation. When this happens, the performance of your program will be hit *drastically*, but it won't fail.

NOTE   As a famous computer scientist, Leon Bambrick, once said: "There are only two hard things in computer science: cache invalidation, naming things, and off-by-one errors."

So, while using Goroutines never guarantees to make your code faster—in fact, in a majority of cases, single-threaded code is faster than naively multi-threaded code—it can help if you engineer the interactions between threads in just the right way.

In Go, this work of engineering the interactions between threads becomes a lot easier because the unique nature of Goroutines affords you the ability to run more concurrent functions at the same time without worrying about as much overhead.

In order to modify the previous proof of work implementation, all we need to do is rewrite the "pow" function in order to support using Goroutines for evaluating certain sequences of bytes. So, you can go ahead and copy your previous code file, remove that function, and start rewriting it.

First, let's discuss the logic of the application. Remember, in this app, we don't need the different Goroutines to interact with each other. We just need to make it so that the Goroutines are able to figure out which string they're trying to find a match for, and then have a way for the Goroutines to give us a solution if they find one. We also want a way for the Goroutines to shut down gracefully once a solution is found. Finally, we need one more thing: a way for us to count how many hashes have been tried in total across all Goroutines. This is important for benchmarking purposes, because we want to know how performant this implementation is against the single-threaded version.

For this, it's important we start off by knowing how many CPU cores the machine we're running the program on actually has. Remember, a computer can only run that many operations at the same moment in time. For example, if a computer has eight CPU cores, it can usually only run eight streams of instructions (threads) simultaneously without having to context switch. We say usually because some CPUs, such as Intel and AMD CPUs, support "simultaneous multithreading (SMT)," enabling two threads to run on each core mostly "at the same time." Some other CPU architectures, such as IBM's Power, can actually be configured through the operating system to support one, two, four, or even eight threads per core—even if each thread becomes slower because fewer core resources are allocated to the thread.

Let's start implementing the function like so:

```go
func pow(prefix string, bitLength int) {
    start := time.Now()
    hash := []int{}
    totalHashesProcessed := 0

    numberOfCPU := runtime.NumCPU()
    closeChan := make(chan int, 1)
    solutionChan := make(chan []byte, 1)
```

This function has the same name, pow, as the previous one. It also takes the same arguments, the prefix string, which is the string we want to find a valid random sequence for, as well as the bit length, which is how many consecutive zeros the resulting hash's bits need to start with to count as a solution to the problem.

Within the function, we start off by determining the current time, so we can find out how much time we spent doing the actual proof of work operation. Then, we create an array of integers called "hash." This array is going to contain the number of hashes that each Goroutine processed. This means that, in total, this array will contain the same number of elements as there are Goroutines. Alongside this array there's also the "totalHashesProcessed" integer, which will store the sum of this array in a future step.

At the same time, we ask Go how many CPUs (CPU cores) the machine we're running on actually has. This way we can spawn the same number of Goroutines as we have CPU cores on the machine. If you spawn too many Goroutines for the number of cores on your machine, you may end up spending more time on the context switching between Goroutines than you actually do on the work within them. While the overhead associated with Goroutines, like that of context switching, may be a lot *less* than threads, it's still not free.

Then, we initialize two channels: closeChan and solutionChan. closeChan will help us close all the Goroutines once a solution has been found, and solutionChan will help us communicate that solution from the Goroutine that found the solution back to the main Goroutine.

Now let's implement the meat of the function. This is a bigger chunk of code, but we'll deconstruct it in just a moment:

```
for idx := 0; idx < numberOfCPU; idx++ {
    hash = append(hash, 0)
    go func(hashIndex int) {
        seed := uint64(time.Now().Local().UnixNano())
        randomBytes := make([]byte, 20)
        randomBytes = append([]byte(prefix), randomBytes...)
        for {
            select {
            case <-closeChan:
                closeChan <- 1
                return
            case <-time.After(time.Nanosecond):
                count := 0
                for count < 5000 {
                    count++
                    seed = RandomString(randomBytes,
                        ↪ len(prefix), seed)
                    if Hash(randomBytes, bitLength) {
                        hash[hashIndex] += count
                        solutionChan <- randomBytes
                        closeChan <- 1
                        return
                    }
                }
                hash[hashIndex] += count
            }
        }
    }(idx)
}
```

Looks fun, right?! You may already have deciphered some high-level understanding of the code while looking through it, but let's dissect it now. Starting, of course, with the loop:

```
for idx := 0; idx < numberOfCPU; idx++ {
    hash = append(hash, 0)
```

As mentioned, we want to spawn the same number of Goroutines as there are cores on this machine to handle those Goroutines. Our application is very special in that it's very CPU-bound, not I/O-bound or memory-bound, meaning that the bottleneck to high performance in this case is not having more memory bandwidth or faster I/O. Instead, you need a faster processor. When applications are *not* CPU-bound, it may make more sense to spawn more Goroutines than there are cores, because while one Goroutine is waiting on the CPU for something else to complete, another one can begin to run.

Because we want to spawn only as many Goroutines as we have cores, we use a `for` loop to iterate a number from 0 to 1 less than the number of cores we have. At each iteration, we append a 0 to the hash array. Remember, the hash array contains the number of hashes each Goroutine has processed.

Next, in the loop, we spawn the Goroutine itself:

```
go func(hashIndex int) {
    seed := uint64(time.Now().Local().UnixNano())
    randomBytes := make([]byte, 20)
    randomBytes = append([]byte(prefix), randomBytes...)
```

The function this Goroutine invokes only takes a single argument, and that is its index; using this index (called `hashIndex` in the code), it can log into the hash array the number of hashes it has processed.

Within this function, we create a new random seed by getting the local Unix Epoch timestamp in nanoseconds. This seed is converted to a 64-bit unsigned integer, which is compatible with the custom random number generator we defined in the last chapter.

Then, we run a little bit of what may look like "funny" logic if you come from languages such as C, Python, Java, or Swift. Essentially, we start by allocating a new array of 20 bytes. These are the 20 bytes in memory that will contain the actual "random" ending to the string, which may give us the solution to our proof of work problem. Then, we insert the prefix string (represented as an array of bytes) in the *beginning* of this new array.

We insert it in the beginning by using the append function. The append function supports *variadic arguments*, which, if you remember from Chapter 2, means it can take any number of arguments of the same type after the first argument. The first argument must be an array of some type, and the rest of the variadic arguments must be of the same type as the elements within the array.

By feeding the `prefix` as the first argument into the `append` function, and then feeding all of the 20 elements in `randomBytes` as 20 arguments after that, we end up essentially creating a new array where the memory looks like this:

```
[bytes of the prefix][bytes we initialized when we allocated
randomBytes]
```

If you're wondering how the last line in that snippet manages to send 20 parameters by only passing 1, it's using the . . . operator. This will "unfurl" the array into 20 separate arguments at runtime, making it look to the callee function as if you had manually passed those 20 arguments.

Next, within the function that the Goroutine invokes, we need an infinite loop that can actually do our processing. This loop has a `select` statement within it because, of course, we want to know when `closeChan` tells us to exit. Here's how we do that:

```
for {
    select {
    case <-closeChan:
        closeChan <- 1
        return
```

This code may look simple (and it is!), but it packs a really interesting piece of logic. Look at the case for receiving from `closeChan`: Why do we put a number *back* into the channel after we receive a number from it? Well, first things first. Recall that, regardless of the fact that we put a number back into the channel, we *do* return from the function if we get a value from this function.

So why do we end up putting a number back in the channel? It's because when we do this, we only need the Goroutine that finds the solution to insert *a single number* into this channel, and *all* of the Goroutines will exit. That's because technically only one Goroutine will pick up the number the solution Goroutine sends, but before that next Goroutine exits, it will tell *another* Goroutine to exit by sending that message back into the channel.

The final Goroutine will send another message to the channel and then exit, but because there are no more Goroutines to pick up the message, that's just going to stay there, which doesn't really matter to us.

```
case <-time.After(time.Nanosecond):
    count := 0
    for count < 5000 {
        count++
        seed = RandomString(randomBytes, len(prefix),seed)
        if Hash(randomBytes, bitLength) {
            hash[hashIndex] += count
            solutionChan <- randomBytes
            closeChan <- 1
            return
        }
    }
    hash[hashIndex] += count
}
```

This next case in the select statement is where the main work actually happens. Before we can discuss the work, we think it's important to discuss what's more than likely the elephant in the room: Why is the case waiting for time.After(time.Nanosecond)?

Again, if you go back a couple of pages, you'll recall we mentioned that a select statement allows you to wait for multiple blocking operations to unblock at the same time. time.After is a blocking operation, and so is waiting for a value on the exit channel. We just need *some* kind of blocking operation here so it is eligible to be part of the select statement, in order to run this code immediately if the exit channel doesn't have any information to receive.

Because we're only asking Go to run this code after a nanosecond, it's essentially a noop ("no-op" or "no operation"), but because we also ask Go to evaluate the exitChan wait first, if there is a value to receive there, we won't end up running this code.

Within the actual case's code, the operations we execute are similar to the previous, raw "pow" function that didn't use Goroutines. Except, instead of running them a single time, we run them 5000 times. This is so we run blocks of 5000 checks in between waiting for the exit channel. If we were to only do a single check before waiting for the channel again, we wouldn't be doing enough work to justify the extra overhead of all these channels, the switch statement, and the Goroutine.

Specifically, we have a for loop that iterates 5000 times, and each time, generates a new random string to append to the prefix. It then feeds the complete string (prefix + randomized portion) into the Hash function, which will tell us whether or not we met the proof of work condition.

A bit of a change from the last chapter comes in what we do if we end up finding a solution. Once we find a random string that meets our requirements, we add the "count" variable (which is how many of the 5000 iterations we've completed locally) to the Goroutine's dedicated element in the hash array. Then, we send the solution to the solution channel and start the exit cycle by feeding a single number into the exit channel. We also return from the function, breaking the loop, and killing this Goroutine.

If we don't find a solution in these 5000 iterations, we increment this Goroutine's dedicated element in the hash array by 5000 and iterate once again through the larger `for` loop, which runs the `select` statement and, by extension, potentially this code, once again.

```
        }
    }(idx)
}
fmt.Println(<-solutionChan)
for _, v := range hash {
    totalHashesProcessed += v
}
end := time.Now()
```

This next section of the code closes the `for` loop inside the Goroutine, closes the function, invokes the function with the index variable from the array, and then closes the `for` loop that's responsible for spawning the Goroutines.

Then, we wait for some output on the solution channel, and once we get it, we print it out. Then, we calculate the number of hashes we processed in *total* across all Goroutines. Finally, we find the end time, which we use in the following code to print out some statistics of the run:

```
fmt.Println("time:", end.Sub(start).Seconds())
fmt.Println("processed",
    ↪ humanize.Comma(int64(totalHashesProcessed)))
fmt.Printf("processed/sec: %s\n",
    ↪ humanize.Comma(int64(float64(totalHashesProcessed)/
    ↪ end.Sub(start).Seconds())))
}
```

And, just like that, you've successfully built your first real application that uses the power of Goroutines!

This is what your code should look like now:

CODE LISTING 5.6 The new, improved, Goroutine-powered pow function

```go
func pow(prefix string, bitLength int) {
    start := time.Now()
    hash := []int{}
    totalHashesProcessed := 0

    numberOfCPU := runtime.NumCPU()
    closeChan := make(chan int, 1)
    solutionChan := make(chan []byte, 1)
    for idx := 0; idx < numberOfCPU; idx++ {
        hash = append(hash, 0)
        //pass in idx to ensure it stay the same as idx can
        //change value
        go func(hashIndex int) {
            seed := uint64(time.Now().Local().UnixNano())
            randomBytes := make([]byte, 20)
            randomBytes = append([]byte(prefix),
                ↪ randomBytes...)
            for {
                select {
                case <-closeChan:
                    closeChan <- 1
                    return
                case <-time.After(time.Nanosecond):
                    count := 0
                    for count < 5000 {
                        count++
                        seed = RandomString(randomBytes,
                            ↪ len(prefix), seed)
                        if Hash(randomBytes, bitLength) {
                            hash[hashIndex] += count
                            solutionChan <- randomBytes
```

```
                    closeChan <- 1
                    return
                }
            }
            hash[hashIndex] += count
        }
    }
}(idx)
}
<-solutionChan
for _, v := range hash {
    totalHashesProcessed += v
}
end := time.Now()
fmt.Println("time:", end.Sub(start).Seconds())
fmt.Println("processed",
    ↪ humanize.Comma(int64(totalHashesProcessed)))
fmt.Printf("processed/sec: %s\n", humanize.Comma(int64
    ↪ (float64(totalHashesProcessed)/end.Sub(start)
    ↪ .Seconds()))))
}
```

Because this function has the same signature, you should be able to keep your main function exactly the same, and compile and run this code. In Chapter 4, we mentioned that on a MacBook Pro, this code reached around 4 million hashes per second.

Well, get ready, because this code reaches nearly *30 million* hashes per second on the same machine. That's around a 7.5× speedup without a massive amount of effort. And, what's great is that this machine runs 8 cores on an Intel i9, so we are reaching ~8× speedup, which is what we expect.

Remember that because of the overhead of spawning, shutting down, and checking the exit channel, scaling is not *perfectly* linear, but it is still really good, especially because, in the grand scheme of software development, we put next to no effort into making it this fast. In a way, you've learned to leverage the processing power of your CPU that otherwise goes unused. By unlocking this capability using Go, you can make your applications scalable and fast.

In this chapter, we've taken a look at what makes Go *really* special: Goroutines, a feature that is literally named after the language! However, while Go does a lot of things really well, there are some things it simply cannot do as nicely as other languages. This isn't a problem though, because Go is a compiled language, and because of the compile process, it's relatively easy to call code from other compiled languages, too! So, in the next chapter, let's take a look at how you can call code from languages such as C and Swift in Go, and why you might want to do so in the first place.

Exercises

1. What is concurrency? How does the concept of threads relate to concurrency?

2. What is the relation between processes, threads, Goroutines, and functions?

3. What are channels and how do they work? Think of and implement a small use case that demonstrates what channels are used for.

4. What is a semaphore? How does a semaphore prevent deadlocks in your code?

5. What kind of channel can be used in Go to replicate a semaphore's functionality?

6. From a syntax perspective, how are a `switch` and `select` statement similar?

7. Why does the proof of work application scale so well to nearly 8× the performance when run on 8 physical cores?

8. What is the most common source of overhead when multithreading applications?

9. How can you get multiple Goroutines to automatically kill themselves with only a single manual channel insert operation?

10. How does variable capturing differ in Go from most languages, and what can be an unintended side effect of this?

Interoperability

Welcome to Chapter 6! In this chapter, we'll take the concept of "reusing code" one step further. When we've finished, you'll not only be able to import external codebases written in Go (known as modules), you'll also be able to import other codebases *written in other languages*!

Once you've finished this chapter, you'll be able to answer the following questions:

- In what format do operating systems understand the binaries generated by compilers?
- What is a language's "runtime"? What effect does it have on the programming experience and compiled code?
- How can multiple programs on the same operating system share the same basic pre-compiled binary code?
- Why is inter-language operability important?
- How can you call C and Swift code from Go using shared libraries?

One principle of computer science is to not reinvent the wheel. This means that if something already exists, then it's usually best not to have to rewrite it. The ideology of the principle is that if someone has already built something, they likely had a purpose for it, and have put some level of thought into how best to solve the problem.

In today's world, where open source software is more prevalent than ever, this principle is even more valuable. This is because, even for tasks that may seem simple on the surface, literally *thousands* of people may have put effort into architecting an effective solution. As an individual developer, it's highly unlikely that your hand-rolled solution could beat the solution that many other developers have poured time and energy into. However, everyone has their own unique perspective to solving problems. So, if you do find an inefficiency, or something you'd like to change, open source software lets you do exactly that.

Why Is Interoperability Important?

You may be thinking: "Didn't we just cover this topic in the last chapter?" Yes, we did. However, we covered using libraries of code written in Go. This is incredibly useful and is what you'll be using in the majority of cases. But not all code in the world is written in Go—in fact, much of it isn't. Every programmer has their own favorite programming language, and every task calls for using a different language. How, then, can you use other people's code in your own Go programs, if that code isn't written in Go?

This is a problem that almost every programming language faces in some way, and every language deals with it in a different way depending on its architecture. Python, for example, is not a compiled language; it's an interpreted language. This means that when you run a python program, your code is not compiled down to machine code. Instead, an interpreter process is actively looking at your code and running it through that process.

Then, there's also the JIT (Just In Time)–compiled languages such as Java. Java compiles your code beforehand, but not to machine code. Instead, your code is compiled to Java Bytecode, which is a higher-level instruction set than the one your CPU uses. When you run the Bytecode, the Java Virtual Machine interprets those low-level instructions in a similar manner as Python. This method usually achieves much better performance than a purely interpreted language such as Python, because the code being interpreted can be compiled on the fly to native CPU code and cached for when the function is called in the future.

Finally, there's the bracket that languages such as Go, C, C++, Swift, and Rust fall under: AOT (Ahead Of Time)–compiled. These languages compile your code directly to native CPU code before you can run the program. Usually, these languages can best optimize your code to run as fast as possible. Plus, because of the way they're architected, they all output a binary executable that your operating system already knows how to run without any additional software.

On macOS, the compilers need to generate Mach-O files. On Linux, they need to generate DWARF files. On Windows, they need to generate .exe files. Each file format has its own standard of arranging the instructions in binary, as well as header information to help the operating system understand how the code is intended to be run.

For most AOT-compiled programming languages, such as C, C++, Swift, and Rust, this is very straightforward. There is an "entry point" or "main" function that the operating system will call into, this function will execute whatever code it needs to execute, and then once the main function exits, the program exits.

Sometimes, your own code can call out to other pre-compiled code on the same system. For example, when you call `malloc(1)` for memory allocation, your own code doesn't contain the definition of the malloc function. Instead, that's a function provided by the C library on your operating system. So, the binary your compiler outputs also gives the operating system instructions on what "libraries" are "linked" to this binary, which your code needs to access.

These kinds of libraries—ones that you promise to your compiler exist at compile time, and that are loaded at runtime—are called "shared libraries." That's because multiple programs can share that same library. So, instead of having every single camera application shipped for an iPhone contain a copy of the library that handles dealing with the camera, the operating system contains a single copy, and that copy is loaded up dynamically when you run an executable that needs it.

On different platforms, these libraries are referred to with different file formats. On Linux, it's ".so" (Shared Object); on macOS, it's ".dylib" (Dynamic Library); and on Windows, it's ".dll" (Dynamically Linked Library). They all have the same basic purpose. Some systems handle these more or less elegantly than others—thus, the occasional "DLL Hell" situation on Windows.

The knowledge of shared libraries brings us to the way we can share code across programming languages. Because AOT-compiled languages translate their high-level code to the same basic format of machine code—even shared libraries—we can just compile some arbitrary code to a shared library. Then, from another programming language, we can take the compiled code and link it against the aforementioned library, enabling us to call the code from that other language.

As you can likely tell, there are limitations with this technique. Mainly, what if the language we want to call is interpreted or JIT-compiled? Well, this definitely introduces a couple of challenges, but in that case the solution varies case by case. For Python, you can link against the interpreter's shared library, and interpret the Python program manually. With Java, you have a similar capability.

In this chapter, we won't explore calling interpreted languages. Instead, we'll deal with AOT-compiled languages only. However, there is one more limitation, and it's one you may have missed if you're not familiar with the internal operation of the Go compiler. This limitation is that not every AOT-compiled language follows the same format of machine code!

Well, from the operating system's perspective, "in theory," yes, each executable follows the same format, regardless of the language it was compiled from. However, in practice, each language usually adds its own set of tricks to its binaries to get them to achieve some kind of special behavior. This added code is called the language's "runtime," and usually, the more "runtime" the language has, the more annoying it is to call its code from other languages, and vice versa.

This is because the code the compiler ends up outputting isn't exactly the code you programmed in. You see, that compiled machine code was built with certain assumptions in mind—assumptions that only hold up if the compiled code is called from other code compiled with the same compiler. Think about Swift, for example. When you pass a reference to a class, Swift's ARC increments a reference counter to ensure that it can keep track of when memory needs to be freed. If you pass that reference as a pointer to some C code, the C code doesn't own the pointer and can't ever be sure that that memory is really safe to access or mutate.

And, as we've already established in the previous chapters, Go has a *very big runtime*, much larger than that of C, Swift, or Rust. This is because Go needs to deal with Goroutines, and all the baggage they add to the fray. So, when you compile a Go program to assembly, you can see that the main function you define isn't even the entry point of the program. The entry point of a Go program is actually inside a shared library linked against it, which contains the majority of the backend of the Go runtime. That shared library then initializes the runtime, gets some internals ready and set up, and then calls your main function. When the program exits, it also of course does so through this library.

Therefore, you cannot just "link the Go program against a shared library and call those functions." Calling out to a shared library in Go actually incurs a relatively large performance penalty, at least when compared to other languages. This penalty can almost always be nullified by using the Gccgo compiler instead of Gc.

Despite the penalty, it's usually worth it to call code from other languages, especially when another language can do a certain task fundamentally better than Go but you'd like to access the result of the computation from your Go code.

To demonstrate this kind of situation, in this chapter we'll explore multiple tasks and ways of interoperating between Go and other AOT-compiled programming languages.

Interoperating with C Code

Let's start with a simple "Hello, World!" example of calling C code from Go. First, let's see what the pure C version of this application would look like:

CODE LISTING 6.1   A simple "Hello World" application in C

```c
#include <stdio.h>
int main(){
    printf("Hello, World!\n");
    return 0;
}
```

The only essential lines are lines 1 and 3. These lines tell the compiler to:

1. Take the "stdio.h" file from the OS and paste its contents into this file.
    a. This file contains a lot of function declarations (not definitions) that deal with input and output. However, because the header only tells our code *which* functions it can call, not *how* they work, you still need to link this code against the library that has those

functions. The stdio.h file doesn't require this manually because your binary is linked to the standard library by default.

2.   Call the "printf" function and pass it a pointer to a character buffer containing the "Hello, World!\n" string.

The other three lines of code are just ceremony to get the C compiler to compile and run this code.

In order to call this kind of code from Go, we need to make a few changes. First, remember that the entry point of the program is not from C, but from Go. This means we cannot define a main function in the C code. Instead, we need to have another function with any name of our choosing. Also, because this won't be the main function, and because it only needs to print and doesn't need to return anything, the return type can be "void", which in C means "nothing."

Therefore, the C code can be simplified to the following:

CODE LISTING 6.2   A "Hello World" function in C

```c
#include <stdio.h>
void printHelloWorld() {
    printf("Hello, World!\n");
}
```

This C code will, by default, not compile and run with any C compiler, because it doesn't contain the main function required for an executable. However, if you were to compile with the following flags (assuming you are using the Clang or GCC compilers):

```
-shared -fPIC -o libhelloworld.so
```

You should get a file called "libhelloworld.so", which contains the preceding code, compiled to machine code, in shared library format. Make sure to change the file extension based off of the OS you're running on.

As we covered previously, this shared library can be called from your Go code. This would be using another Go library known as cgo. With cgo, not only can you compile Go code against certain C headers and link against precompiled C code, you can even give it raw C code and have it compile that code *into* the final executable that Go produces. This makes it so you don't need a shared library in the first place.

That's what we'll explore first. In order to use cgo, all you need to do is import the "C" package like so, with all the C code you'd like to run as comments before the import:

```
package main

//#include <stdio.h>
//void printHelloWorld() {
//    printf("Hello, World!\n");
//}
import "C"
```

This will make Go import the C module, then it'll look at the comments that came before it, and treat that as C code that it needs to compile into the Go executable.

Now, in order to call that C function, you can simply treat it as any other function in the C module:

CODE LISTING 6.3   Calling the C "Hello World" function from Go

```
package main

//#include <stdio.h>
//void printHelloWorld() {
//    printf("Hello, World!\n");
//}
import "C"

func main() {
    C.printHelloWorld()
}
```

If you were to compile and run that code, you should get an executable that prints hello world via C instead of Go.

Now that you understand how to call very simple C code from Go without needing a library, let's explore a slightly more complex example. Specifically, what if you want to pass variables to C or use the return values from the functions?

Our first hurdle is the translation of types from C to Go, and vice versa. The cgo module contains a bunch of types, such as C.int (32-bit signed integer), C.uint (32-bit unsigned integer), C.long (64-bit signed integer), and more.

One of the most difficult types to convert is the string. This is because strings aren't technically one piece of data. They're a list—an array—of multiple independent pieces of data, which we call characters. What's worse is that what we consider, as humans, to be a single "character" may not be represented by a single byte, as in the case of emoji or other Unicode characters.

In C, strings are represented as raw arrays of 8-bit signed integers, called "char*" (or "pointer to a character buffer"). The last, terminating value in these strings is usually a byte with all bits set to OFF (to 0), which acts as a "null terminator"; thus, the name "null terminated strings." This tells us when the string buffer actually ends so we don't need to keep track of the length alongside the buffer forever.

In Go, translating a Go string to a C string is quite easy with the help of the CString function from cgo. However, keep in mind that this makes a copy of the memory backing the string, which can be bad in some cases if the code is performance sensitive. This is a price you have to pay when you're interoperating between languages that are high level and have the baggage of a runtime.

This also means that after you use the C string that you make, you are responsible for freeing that memory and making sure it doesn't live longer than it needs to. This is called a *memory leak*. Go's Garbage Collector cannot deal with this pointer for you, because you created it and are responsible for freeing the buffer of memory it points to.

The following is the signature for the CString function:

```
func C.CString(string) *C.char
```

Now let's use a slightly more complex C function in Go. Specifically, let's define a function that simply takes a null terminated string and prints it out. Here's all the code you'll need to do that:

CODE LISTING 6.4 Passing character pointers to C from Go

```
package main

//#include <stdio.h>
//#include <stdlib.h>
```

```
//void printString(char* str) {
//    printf("%s\n", str);
//}
import "C"
import "unsafe"

func main() {
    a := C.CString("This is from Golang")
    C.printString(a)
    C.free(unsafe.Pointer(a))
}
```

As you can see, the logic is quite simple. First, we implement the C code in comments before the import of the cgo module. The C function simply takes a character pointer and passes that to the `printf` function, along with a format string telling it to print the character pointer followed by a new line.

However, you need to do a bit more work in the main function. The first thing is to convert the Go string to a C string, which is done by `C.CString("This is from Golang")`. Then, pass it as an argument to the C function, just as you would to any other Go function.

Because you're in control of the memory manually, you also need to free the buffer yourself. Remember, though, that the C string is a character pointer, not just a generic pointer. The C free function also technically expects a generic pointer with no type information attached to it. Because Go is strict with types, you need to use the Pointer function in the unsafe module to erase the type associated with the C string. Then, you can pass it to the free function, and you're good to go!

One more thing—the definition of the free function is the reason we had to also include `stdlib.h` in the C code this time alongside `stdio.h`. Without that header, there would be no way for us to free the memory we allocated.

If you were to run this code, you should see "This is from Golang" printed to your terminal. Thus, the string is internally passed from Go to C, printed, and then the memory is freed.

Taking return values that come from C functions and translating them back into types that Go can understand is also pretty simple. For example, if you were to *return* a string (character pointer), you can convert it to a Go string like so:

CODE LISTING 6.5   Converting a C string to a Go string

```
package main

//#include <stdio.h>
//char *getName(int idx) {
//    if (idx == 1)
//         return "Tanmay Bakshi";
//    if (idx == 2)
//         return "Baheer Kamal";
//    return "Invalid index";
//}
import "C"
import (
    "fmt"
)

func main() {
    cstr := C.getName(C.int(2))
    fmt.Println(C.GoString(cstr))
}
```

In this case, you don't need to free the character pointer returned from C because that value was never allocated on the heap in the first place. It was only stored on the stack. However, if the buffer originated as the result of a call to `malloc(1)`, then the `free` is required.

One last thing you should know about—you are allowed to instruct the C compiler to compile the code you insert in your comments however you'd like it to. So, for example, if you want to pass a specific compile flag, you can do so using the `CFLAGS` parameter.

Let's say you're writing a function to tell whether or not a number is "ugly," meaning its prime factors include only 2, 3, and 5. These are the first 20 ugly numbers:

```
1, 2, 3, 4, 5, 6, 8, 9, 10, 12, 15, 16, 18, 20, 24, 25, 27, 30, 32, 36
```

The following could be some C code to achieve this:

CODE LISTING 6.6   Code to get the nth ugly number in C

```c
#include <stdio.h>

int numberIsUgly(int x) {
    while (x > 1) {
        int y = x;
        while (y % 2 == 0)
            y /= 2;
        while (y % 3 == 0)
            y /= 3;
        while (y % 5 == 0)
            y /= 5;
        if (x == y)
            return 0;
        x = y;
    }
    return 1;
}
=
void getNthUglyNumber(int n) {
    int i = 0;
    int j = 0;
    while (j < n) {
        i++;
        if (numberIsUgly(i)) {
            j++;
        }
    }
    printf("%d\n", i);
}
```

Using the getNthUglyNumber function, you can get the nth ugly number. For example, the 10th ugly number is 12. The 20th is 36. This algorithm can definitely be improved, but let's roll with this one for now.

Write the following Go code to wrap the C code:

CODE LISTING 6.7 Calling the ugly number code from Go

```
package main

//#cgo CFLAGS: -O0
//#include <stdio.h>
//
//int numberIsUgly(int x) {
//    while (x > 1) {
//        int y = x;
//        while (y % 2 == 0)
//            y /= 2;
//        while (y % 3 == 0)
//            y /= 3;
//        while (y % 5 == 0)
//            y /= 5;
//        if (x == y)
//            return 0;
//        x = y;
//    }
//    return 1;
//}
//
//void getNthUglyNumber(int n) {
//    int i = 0;
//    int j = 0;
//    while (j < n) {
//        i++;
//        if (numberIsUgly(i)) {
//            j++;
//        }
//    }
//    printf("%d\n", i);
//}
```

```
import "C"

func main() {
    C.getNthUglyNumber(C.int(1000))
}
```

Notice that there is one significant difference between this code and the code we've written previously using cgo. It's in the first line of C code: We're telling cgo to pass a flag "-O0" to the C compiler using the CFLAGS variable. This flag tells the C compiler "Don't optimize the code at all! Do as much of a direct code to assembly translation as possible."

When you run this code, the process of finding the 1000th ugly number will take about 1.7 seconds on a MacBook Pro with an Intel i9. That's a long time! If, however, you were to change the flag at the top of the code from O0 to Ofast, like so:

```
//#cgo CFLAGS: -Ofast
```

You should see that the code now only takes about 0.4 seconds to execute on the same machine. Compiler optimizations are powerful, and using the CFLAGS variable, you can pass these and any other GCC flag to the compiler.

By the same logic, you can also pass flags to the "linker," which is responsible for telling the operating system which shared libraries your executable needs to call code from (or "is linked to").

To test this out, write the following C code into a file named "factorial.c":

```
int factorial(int x) {
    if (x == 1)
        return x;
    return factorial(x - 1) * x;
}
```

You don't need any includes or a main function, because we will simply compile this code, which calls no other function but the one that it defines, into a shared library. Use the following compile command to do this:

```
clang factorial.c -shared -fPIC -o libcfactorial.dylib
```

Now you should have a file containing the binary machine code for the factorial function. In order to call this from Go, the following is all you have to do:

CODE LISTING 6.8   Linking against a shared library and calling a C function from Go

```
package main

//#cgo LDFLAGS: -L. -lcfactorial
//int factorial(int);
import "C"
import "fmt"

func main() {
    fmt.Println(C.factorial(C.int(5)))
}
```

There are only two lines of comments for cgo, and this is what they do:

1. The first one tells the linker "look for shared libraries in the directory in which the code is being compiled, and specifically link against a library called 'cfactorial'."

   The real filename of the library is "libcfactorial.dylib", but the linker ignores the "lib" in the beginning and the ".dylib" at the end.

2. The second one says "I don't know how this function works, but I know there is a function somewhere in the program called factorial, it returns a 32-bit signed integer, and takes a single 32-bit signed integer as an argument."

   This function's actual machine code is already compiled in the library that we built in the previous step.

To be even more clear, remember that when you now compile your Go code, you will *not* be compiling the factorial function. You already did that compilation when you ran the clang command manually. Instead, you're now telling Go to link against the precompiled version of the factorial function.

When you build and run the Go code, you should see it print out the factorial of 5:120.

You're now in the know of all of the basics when it comes to calling C from Go. It's time to move on to something more complex: calling Swift from Go, and actually building a practical application that makes sense in the real world.

Interoperating with Swift

First off, why would we ever want to call Swift from Go?

Remember, the Go compiler is really powerful. However, Go is fine with making a very specific tradeoff, because it's deemed beneficial for the kinds of codebases that are written in Go: the compilation speed versus runtime speed tradeoff. Go is okay being *really* fast at compiling your code, even if that means the compiled code itself may be a bit slower, or if the language is a bit less expressive because of it.

Mostly, this doesn't really matter, because the Go optimizer is really good anyway. But there are some optimizations you just don't get, and that could be very beneficial for specific operations. For example, tail call optimizations don't exist in Go! Tail call optimizations can sometimes make recursive algorithms significantly more efficient by making it so the recursive call doesn't need to return back through the deep recursive call stack and can return back to the original caller instead.

For this kind of code, you might think that writing in a language such as C becomes important. And while this may be true, sometimes you don't need the level of control that C gives you; it can be a burden. This is why Swift, and other similar languages, exist.

However, there are also some tasks that *Go* is really good at, such as concurrency. Or, maybe there's a specific module that someone wrote in Go that makes a certain aspect of your application a lot easier to code. Sometimes, you need to leverage the capabilities of Go in the same application that requires code to be written in another language. This is when calling another high-level language, such as Swift, in Go can be useful.

To demonstrate a use case where this is true, we'll implement an application that uses a special "AI" to play Tetris* automatically, and to visualize the output of the Tetris game in the command line. To understand how this requires an interop between Go and Swift, we first need to understand the game of Tetris and then dive deeper into the AI we'll build.

You've almost certainly played or seen someone play Tetris before, but in brief, Tetris is the classic game of taking falling bricks and manipulating the way they fall from the top of the screen to the bottom such that the bricks can make "lines" across the board. When you build a line, that row of the board gets cleared and any bricks above it get moved down.

---

* Tetris is a registered trademark of The Tetris Company, LLC.

Figure 6.1 is a sample screenshot of an implementation of the classic Tetris game as someone is playing it. Because the version that we build in this chapter runs in the command line, it will look different.

FIGURE 6.1   Playing the classic Tetris game by The Tetris Company, LLC

The specific strategy taken by the AI that we'll build is to:

1. Look at the block that is falling and determine what all the possible final states are of that block on top of the other blocks—regardless of whether they're good or not.
2. Use many weighted heuristics to score each possible end state to find out which one is the "best," in that it enables us to have the greatest likelihood to maximize the score.
3. Use a pathfinding algorithm to find out how to get to that ending position.
4. Make the moves to get to that end state.

In this way, with the right set of heuristics and weights for each heuristic, we can get an AI that's really good at playing Tetris.

The following are the heuristics we'll implement:

- **Number of holes**—Gaps in the blocks that can't be filled until the lines above have been cleared
- **Number of open holes**—Gaps in the blocks that are not covered by other blocks
- **Blocks above holes**—Number of blocks above the holes
- **Maximum line height**—The peak Y position of a block on the board
- **Last block added height**—The peak Y position of the block that we're adding in this move
- **Number of pillars**—Columns that require a vertical line piece to fill
- **Blocks in rightmost lane**—The peak Y position blocks in the rightmost column
- **Bumpiness**—Cumulative difference of the peak Y positions between each neighboring column pair

The way a certain state's "cost" will be determined is by getting the values of all these heuristics, multiplying them by a weight, and summing the weighted heuristics. The lower the cost, the better the state is for the game as a whole.

This brings us to the next challenge: How in the world can a human find the right weights to use?

One way of doing this is to use your intuition. For example, we know bumpiness is bad, and so is the number of holes, but open holes are better than blocked holes, and we want to minimize the line height, and so on, and to guesstimate some numbers off of that.

Estimation in this way is simply not scalable because we don't understand the effects of how each heuristic interacts with every other heuristic. This is why we need a mathematical, automated way of finding the weights.

Specifically, the solution implemented in the code is known as Covariance Matrix Adaptation Evolutionary Strategy (CMA-ES). This is an algorithm commonly used when you have a non-differentiable function, or if the derivative doesn't make sense to optimize against. That is exactly the case in this scenario—it doesn't make sense to differentiate through the Tetris board, even if you could. Therefore, we use CMA-ES to find the best weights possible.

The algorithm starts with each weight being just 1—meaning we just sum together the outputs from the heuristics. As you can imagine, this set of weights results in chaotic behavior that doesn't make much sense. However, through a trial-and-error optimization method, CMA-ES quickly finds a solution that can get over *4,000,000* lines cleared! Plus, the optimization only takes under 12 hours to complete.

Something you should keep in mind, however, is that just simulating the game in memory, with no UI rendering, in which the AI got 4,000,000 lines cleared, took over 4 hours. That is a *lot* of time, even if the AI got an insanely high score. However, it's also a *lot* faster than other implementations out there on the internet, mostly because:

- It's written in Swift, so it's compiled and optimized by the best compiler infrastructure in the world.
- No UI rendering occurs, and the game is played entirely in memory.

Also, because the optimization requires you to simulate a full game, the faster we can simulate games, the faster we can find a set of weights that is good, and the more time we can dedicate to the optimization, resulting in even better weights, meaning a better AI.

In order to simulate games faster, you need an implementation of the game logic itself that is fast. This brings us to why the game is implemented in Swift: It needs to be fast. The little optimizations that Go cannot implement for us add up and prevent the final AI from being as competent as it can be. This is a direct example of when better technical infrastructure can enable a better final user experience.

So, now you understand why the AI and game logic is written in Swift. But what about the optimizer itself? It's not called as often as the rest of the game logic, so why not write that in Go? The simple answer: *Don't reinvent the wheel for something so complex, and something that thousands of people have already invested time in.*

A Python package already exists that enables you to optimize with CMA very easily. In fact, it's just a few function calls. But this brings us back to the beginning of the chapter. Didn't we say we wouldn't cover calling interpreted languages such as Python?

Well, yes, we will not dive into the technical details of how the Python interop works. However, Swift also has one more incredible capability: There is a very easy and powerful Python interop layer built into the standard library! So, for example, we can use Python packages alongside our Swift code very easily:

CODE LISTING 6.9   Calling Python code from Swift

```
import PythonKit

let np = Python.import("numpy")

let x = np.array([1, 2, 3])
for i in x {
    print(i)
}
print(x * 2)
```

This is a capability that, unfortunately, does not yet exist in Go. Therefore, the optimization code will also be written in Swift, which will call Python for the real optimization work.

"What will Go do, then?" you ask. Well, Go will be responsible in this case for providing the UI for the application. Because we've already built a GUI in this book, it's time for a CLI, or command-line interface. Go has an excellent library known as "tcell" that is unmatched by anything in the Swift ecosystem. The tcell module provides a very intuitive way to build complex command-line interfaces. Therefore, the main game logic will be handled by Swift, with some parts imported from Python, and all of that will be called from Go.

Isn't it incredible to see how all the code written in these languages can coexist?!

To begin, we need to write the Swift code to actually play the game. You can download this code from the GitHub repo containing all of the code for this book under the "Tetris" folder. The reason you can't just read the code from the book is, well, it's around 700 lines of code—way more than we're sure you want to see in a book!

However, the really important parts of the code are the ones that Go will call. This is what that looks like:

```
var game = Tetris(width: 10, height: 24)

@_cdecl("nextBestMoves")
public func nextBestMoves() -> UnsafeMutablePointer<Int> {
    var nextMoves = game.nextBestMoves()!.0
    nextMoves.insert(nextMoves.count, at: 0)
```

```
        return
            ↪ nextMoves.withUnsafeBufferPointer { ptrToMoves ->
            ↪ UnsafeMutablePointer<Int> in
            let newMemory =
                ↪ UnsafeMutablePointer<Int>.allocate(capacity:
                ↪ nextMoves.count)
            memcpy(newMemory,
                ↪ nextMoves, nextMoves.count *
                ↪ MemoryLayout<Int>.size)
            return newMemory
        }
    }


    @_cdecl("playMove")
    public func playMove(move: Int) {
        switch move {
        case -1:
            game.swapHold()
        case 0:
            game.attemptSpin(clockwise: true)
        case 2:
            game.attemptSpin(clockwise: false)
        case 4:
            game.horizontalMove(left: false)
        case 5:
            game.horizontalMove(left: true)
        case 6:
            game.down()
        default:
            fatalError()
        }
    }


    @_cdecl("renderFrame")
    public func renderFrame() -> UnsafeMutablePointer<Int> {
        var x = [24, 10] + game.render().reduce([], +)
        x.insert(x.count, at: 0)
```

```
    return x.withUnsafeBufferPointer { ptrToMoves ->
        ↪ UnsafeMutablePointer<Int> in
        let newMemory =
            ↪ UnsafeMutablePointer<Int>.allocate(capacity:
            ↪ x.count)
        memcpy(newMemory, x, x.count * MemoryLayout<Int>.size)
        return newMemory
    }
}


@_cdecl("lockGame")
public func lockGame() -> Bool {
    return game.lock()
}


@_cdecl("resetGame")
public func resetGame() {
    game = Tetris(width: 10, height: 24)
}
```

The architecture of the code is simple, and it was designed with the idea that it will be compiled to a shared library. It works by declaring the game state as a global variable and using functions to modify that global state. This architecture can definitely be improved, for example by passing the state back and forth as a pointer instead of it being global. However, for this specific example, it's sufficient to use a global variable.

When the shared library is loaded, the global "game" variable is automatically initialized to a new Tetris board with a default width and height of 10 and 24, respectively.

Then, we have the actual functions that do the work. This is what each one does:

- **nextBestMoves**—This function takes the current board state and runs the whole AI pipeline to determine the next best set of moves. It returns the number of moves and then the moves themselves as an "array" or an "integer pointer," because the return type needs to be C-compatible.
- **playMove**—This function takes an individual move as an integer and runs that move on the game state.

- **renderFrame**—This function takes the current final board (all the pieces that have already landed) and the current piece that is still being finalized (the one in the air) and renders them both onto a single 2D array, which is flattened and returned as an integer pointer (representing an array).
- **lockGame**—This function takes the current piece in the air that has been moved into place via the playMove function and modifies the Tetris game state to move that piece into the final game board, clears any lines that have been formed, makes a new random piece, detects if the game was lost, and returns that as a Boolean.
- **resetGame**—This function resets the global game state by initializing it to a new Tetris structure.

You may notice that each function declaration is preceded by the "@_cdecl" function decorator. This is telling Swift that "this function needs to look and act like a C function in the resulting binary machine code." By doing this, we can call these functions from Go as if they were C functions, regardless of which Swift features they may be using internally.

To compile the Swift code, you can simply run:

```
swiftc TWAI.swift -O -emit-library -o libTWAI.so
```

This is what every flag does:

- `-O`—Use full compiler optimizations and make the code as fast as possible, without using unsafe features.
- `-emit-library`—Don't build an executable binary, build a library that other code can dynamically link against.
- `-o libTWAI.so`—The output filename is "libTWAI.so".

However, installing the Swift for TensorFlow compiler (which is the specific distribution we need) can be a bit tedious. So, to keep the instructions the same for every platform, the GitHub repo containing this example also has a Dockerfile you can build, which will automatically deal with setting up your environment and running the code for you.

Regardless, once the library is built (whether you do it manually or through the Dockerfile), the next step is to implement the library's functionality in Go and actually make it do something. We can now begin coding the Go part of this app!

To start, you'll need to make a few cgo declarations and import a few modules:

```
package main

// #cgo LDFLAGS: -L. -lTWAI
// long *nextBestMoves();
// void playMove(long move);
// long *renderFrame();
// char lockGame();
// void resetGame();
import "C"
import (
    "fmt"
    "os"
    "reflect"
    "time"
    "unsafe"
    "github.com/gdamore/tcell/v2"
)
```

As we analyze this code, also keep in mind that it's important the cgo import is separate from the other imports, because that's the only way it can look at the comments, which contain the C code, that come before it.

The first instruction we give cgo is to tell the linker it needs to link against a library called TWAI, meaning that on Linux the name of the shared library file would be libTWAI.so.

Then, we declare the functions that the shared library exposes, which we're meant to call from the Go code. As you can see, the Swift types have been translated to their C counterparts. The `UnsafeMutablePointer<Int>` type becomes `long *` and Bool becomes `char`.

Apart from that, we also import a couple of modules that are important for the rest of the code, such as Tcell for the CLI.

Now let's build a few helper functions so we can actually call the Swift code from our Go code more easily. These are important especially because we are dealing with buffers and arrays and the like, which need a bit more care.

Let's start with the simpler function. `playNextMove` is where we'll begin:

```
func playNextMove(move int) {
    nextMove := C.long(move)
    C.playMove(nextMove)
}
```

As you can tell, the logic is pretty simple. We take the move as an integer, cast it via `C.long`, and pass it to the function. No pointers, no return values, no messy stuff.

`nextBestMoves` is where things may start to look a little weirder. This is what the helper function looks like:

```go
func nextBestMoves() []int {
    moves := C.nextBestMoves()
    size := int(*moves)
    p := uintptr(unsafe.Pointer(moves)) + unsafe.Sizeof(size)
    sh := &reflect.SliceHeader{Data: p, Len: size, Cap: size}
    return *(*[]int)(unsafe.Pointer(sh))
}
```

This looks scary, but it makes a lot of sense once you dig into it. Let's start with line 3.

To understand what this line does, it's important to understand the output from `nextBestMoves`. As mentioned previously, this function outputs the moves that need to be taken to move the current falling piece to where it should be. This array of moves starts with the number of moves there are to be made. This is so that we know how many elements there are in the buffer. So, in line 3, we simply dereference the beginning of the buffer we got back from the function, and cast that to a Go integer.

After that, we need to find the pointer address where the actual content of the array starts, because we've already looked at the size. Doing this is simple—increment the pointer address we got from the function by the number of bytes we've already looked at. To do this, we cast the pointer to `uintptr`, which we then increment by the size in bytes of the size variable that we've already dereferenced.

We technically already know the size is 8 bytes because it's a "long" or "64-bit integer," but for good programming practice, we still call `Sizeof` to get the size for us dynamically.

Then, we have the pointer address of the buffer itself. To convert this to an array of integers in Go, we use the `SliceHeader` function from the reflect package. This will take the pointer, along with the length of the array itself and the total capacity of the buffer, and convert it to a "slice" or "array." We store the reference to the result that this returns.

Finally, the return of this function may look perplexing at first. But when you break it down, it's pretty simple to understand:

```
return *(*[]int)(unsafe.Pointer(sh))
```

Basically, we tell Go: "Take the 'sh' variable, pass it to the `Pointer` function, take the result, cast it to a pointer to an integer array, then dereference the pointer and return the integer array." You can break it down like so:

```
return → return the following expression's value
      * → dereference the following pointer
    (*[]int) → cast the following value to this type
              (unsafe.Pointer(sh)) → take the result of
                                     Pointer(sh)
```

Finally, there's also the renderFrame function:

```
type Frame struct {
    board []int
    width int
    height int
}

func renderFrame() Frame {
    moves := C.renderFrame()
    size := int(*moves) - 2
    height := uintptr(unsafe.Pointer(moves)) +
        ↪ unsafe.Sizeof(size)
    width := uintptr(unsafe.Pointer(moves)) +
        ↪ unsafe.Sizeof(size)*2
    p := uintptr(unsafe.Pointer(moves)) + unsafe.Sizeof(size)*3
    sh := &reflect.SliceHeader{Data: p, Len: size, Cap: size}

    return Frame{
        board: *(*[]int)(unsafe.Pointer(sh)),
        width: *(*int)(unsafe.Pointer(width)),
        height: *(*int)(unsafe.Pointer(height)),
    }
}
```

Alongside this function, we also declare a new `Frame` structure to store the board as a 1d array, with the `width` and `height` of the board.

Within the function itself, we have a very similar mechanism to `nextBest-Moves`. In this case, while the return value from the Swift `renderFrame` function is just a 1d array, it begins with three integers instead of just one: the size of the total buffer including the other two integers; and the other two integers are the height and then width of the Tetris game board.

`renderFrame` will first calculate the size of the board itself, not including the other two integers, and then will increment the pointer twice, each time by the size of an integer, to get the pointers to the height and width integers as well.

Finally, by incrementing the pointer once more, we get the pointer to the actual array of values, which is then fed through a very similar mechanism to the `nextBestMoves` function.

The Frame is then returned from the function.

And that's all there is to this application! We know that some parts of it—especially the parts that deal with memory, pointers, and buffers—are a bit scary, but the end result really is worth it. If you were to clone the GitHub repo and go into the Tetris folder, then run the following two commands:

```
docker build --tag block_ai:1.0 .
docker run -it block_ai:1.0
```

You should see your terminal turn into a Tetris board automatically playing itself! How great is that?!

The main function is a bit too long to delve into in this book at around 100 lines of code. However, all of this logic is stuff you've already learned and the code is well-commented, so feel free to delve deeper into the code in the repository itself.

Through this example, you should be able to get up and running with interoperability between Go and other languages. You can now apply this knowledge to call code you may have already written in other languages, or may need to write in the future due to specific constraints.

With that, you conclude your journey of diving into the Go programming language. You've not just gotten your toes wet, you've gone for a whole swim. And this is just the beginning! There's still so much more to Go, and so much more to build. You should now be capable of building all kinds of simple and intermediate-level applications using Go, from command-line utilities to apps powered by CLIs and GUIs.

Exercises

1. What does the tcell package enable us to do?
2. Why are interpreted and JIT-compiled languages more difficult to interoperate with?
3. List some scenarios in which cross-language interoperability is critical.
4. What is a language's "runtime" comprised of?
5. What are some common problems when embedding C code in Go?

Index

A

AOT (Ahead Of Time)–compiled
  languages, 157–159
append function, 21
ARC. *See* Automatic Reference Counting
  (ARC)
arguments, 32–34
  variadic, 148
ARM/ARM64, 4
arrays, 20–22
async channels. *See* buffered channels
Automatic Reference Counting (ARC),
  5, 6

B

Bambrick, Leon, 145
binary form, assembly in, 3
bitwise operators, 23–24
blockchain, 113–114
Boolean operators, 23–24
buffered channels, 133, 135, 137
  See also channels

building your own packages, 74–78
built-in packages, 60–70

C

C code, interoperability with, 159–168
C++, 2
capitalization, 15, 17–18
cell automaton, 99–100
cgo library, 160–162
channels, 31, 128, 129–139
  async, 137
  buffered, 133, 135, 137
  and select statements, 139–144
  sync, 137
  unbuffered, 132, 135, 137
  with zero-sized elements, 135–137
char*, 162
compiler, 4–6
  architecture, 3
  building the Go compiler from source,
    11–12
  cross-compiling, 7

compiler (*continued*)
    LLVM, 7
compile-time ownership disambiguation,
        5–6
concurrency, 6–7, 121, 124, 144, 145
    See also channels; threads
conditional branching, 23
    See also if statements; switch
        statements
Conway, John, 99
Conway's Game of Life, 99–112
cores, 6, 124, 145, 146, 148
    and Goroutines, 128
Covariance Matrix Adaptation
        Evolutionary Strategy (CMA-ES),
        171–172
CPU architectures supported, 4
cross-compiling, 7
cryptographic hashing functions,
        112–113
C-style for loops, 30–31
curly braces, 21

D
deadlock, 132–133
defer, 38–40, 54–56
delimiters, 18
design goals, 2–3
Dijkstra's pathfinding algorithm,
        82–99
directory structure, 13

E
errors, 3, 36–37, 51–56

F
for loops, 26–31, 129–130, 140, 141,
        150–151

for-in loops, 26–30
functions, 31–42
    helper functions, 114–118
    passing functions to other functions,
        37–38

G
Game of Life, 99–112
Garbage Collection (GC), 5–6
gc (Go compiler), 11–12
    See also compiler
GCC (GNU C Compiler), 12
gccgo, 12
generics, 40–42
global constant variables, 18–19
global static variables, 16–17
Go modules, 71–74
Gopher (Go mascot), 2
Goroutines, 7, 126–129, 137, 141–146,
        148–149
green threads, 125
    See also threads
Griesemer, Robert, 2

H
hash collision, 113
hashes, 146
"Hello, World!" program, 14–15
    with C code, 159–168
helper functions, 114–118
heuristics, 83, 170–171
hybrid threading, 125–126
    See also threads

I
IBM PowerPC, 4
IBM Z, 4
if statements, 23–25

infinity, 89
installing Go
    building from source, 11–12
    manually installing a precompiled
      binary, 10–11
    using a system package manager,
      10
interfaces, 47–51
interoperability, 156–159
    with C code, 159–168
    with Swift, 169–180

J

Java, 2
Java Bytecode, 156–157
JIT (Just In Time)–compiled languages,
    156–157

K

keywords, 13–14

L

Linux, installing Go, 11
LLVM backend, 7
local variables, 19–20
loops, 26–31

M

main function, 31, 130
make function, 21–22
Martin, Edwin, 100
memory leaks, 162
memory management, 5–6
MIPS, 4
"M:N" model of threading, 125–126
    See also threads
modules, 71–74
multiprocessing, 6

N

native threads, 125, 126, 128, 132
    See also threads
noop, 150
null terminated strings, 162

O

OMDb API, 60–70
operators, 23–24, 35, 37, 43, 45

P

package management systems, 60
packages
    building your own, 74–78
    built-in, 60–70
    overview, 12
    third-party, 71–74
panic, 54–56
pathfinding algorithm, 82–99
Pike, Rob, 2
platforms supported, 3–4
Println function, 15
project structure, 12–14
"Proof of work" application, 112–121,
    144–154
pseudorandom number generators,
    115–117
Python, 60, 156, 172–173

R

random number generators, 115–117
reallocation, 21
recover, 54–56
reference counters, 5
reference cycles, 5
reserving capacity, 22
RISC-V, 4
runMathOp function, 37–38

runtime, 4–6, 158–159
Rust, 6

S

select statements, 139–144, 150
SHA256 hashing function, 113, 117
shared libraries, 157–158, 159, 167–168
`squareIt` function, 129, 131
SSH keys, 75
structures, 42–47
    mutating a structure passed to a
        function as an argument, 44
    mutating a structure through its
        pointer passed to a function, 44
    mutating a structure using a pointer
        receiver, 47
    mutating a structure using a value
        receiver, 46
    using a structure to replace multiple
        return values in a function, 43
Swift, 5, 6
    interoperability with, 169–180
`switch` statements, 23, 25–26
sync channels. *See* unbuffered channels
syntax, 13–14
    capitalization, 15, 17–18
    for Goroutines, 126–127
    and variables, 18

T

tags, 63

Tetris, 169–180
third-party packages, 71–74
Thompson, Ken, 2
threads, 6–7, 124–126, 145–146
    M:N (hybrid), 125
    1:1 (kernel-mode), 125
    N:1 (user-mode), 125
tokens, 13, 129
type cast, 32–33
types, 20

U

unbuffered channels, 132, 135, 137
    See also channels

V

values, 42
`var` keyword, 18
variables, 16–22
    functions and, 34–36
variadic arguments, 148

W

`while` loops, 30–31

X

x86/x86_64, 4

Z

zero-player game, 99